RELIQUIAE DILUVIANAE

RELIQUIÆ DILUVIANÆ

OR,

OBSERVATIONS

ON THE

ORGANIC REMAINS

CONTAINED IN

CAVES, FISSURES, AND DILUVIAL GRAVEL

AND ON

OTHER GEOLOGICAL PHENOMENA

ATTESTING THE ACTION OF AN

UNIVERSAL DELUGE

WILLIAM BUCKLAND

ARNO PRESS

A New York Times Company

New York / 1978

Editorial Supervision: ANDREA HICKS

————◦◦————

Reprint Edition 1978 by Arno Press Inc.

Reprinted from a copy in
  The University of Georgia Library

HISTORY OF GEOLOGY
ISBN for complete set: 0-405-10429-4
See last pages of this volume for titles.

Manufactured in the United States of America

Publisher's Note: Plates 25, 26 and 27
have been reproduced in black and white
in this edition.

————◦◦————

Library of Congress Cataloging in Publication Data

Buckland, William, 1784-1856.
    Reliquiae diluvianae.

    (History of geology)
    Reprint of the 1823 ed. published by J. Murray,
London.
    1. Paleontology--Europe. 2. Caves--Europe.
3. Deluge. I. Title. II. Series.
QE753.B8 1977   551.7'001   77-6510
ISBN 0-405-10433-2

RELIQUIÆ DILUVIANÆ;

OR,

OBSERVATIONS

ON THE

ORGANIC REMAINS

CONTAINED IN

CAVES, FISSURES, AND DILUVIAL GRAVEL,

AND ON

OTHER GEOLOGICAL PHENOMENA,

ATTESTING THE ACTION OF AN

UNIVERSAL DELUGE.

―――――

BY THE REV. WILLIAM BUCKLAND, B. D. F. R. S. F. L. S.

MEMBER OF THE GEOLOGICAL SOCIETY OF LONDON; OF THE IMPERIAL SOCIETIES OF MINERALOGY AND,
NATURAL HISTORY AT PETERSBURG AND MOSCOW; AND OF THE NATURAL HISTORY SOCIETY AT
HALLE; HONORARY MEMBER OF THE AMERICAN GEOLOGICAL SOCIETY; CORRESPONDENT
OF THE MUSEUM OF NATURAL HISTORY OF FRANCE; FELLOW OF C. C. C. AND PROFESSOR
OF MINERALOGY AND GEOLOGY IN THE UNIVERSITY OF OXFORD.

―――――

LONDON:
JOHN MURRAY, ALBEMARLE-STREET.
―――
1823.

LONDON:
PRINTED BY THOMAS DAVISON, WHITEFRIARS.

My Lord,

The investigation which has led me to the present work was begun in obedience to your Lordship's immediate advice; I know not, therefore, to whom I can so fitly dedicate the results of an inquiry, which but for this timely encouragement I might never have undertaken. It has, already, produced conclusions, which throw new light on a period of much obscurity in the physical history of our globe; and, by affording the strongest evidence of an universal deluge, leads us to hope, that it will no longer be asserted, as it has been by high authorities, that geology supplies no proofs of an event in the reality of which the truth of the Mosaic records is so materially involved.

The warm interest your Lordship has been pleased to take in the progress of these later discoveries, which are now published to the world, demands this tribute of my grateful acknowledgment: and I have been long indebted to your Lordship, for the same indulgent notice of my endeavours to call the attention of the University to the subject of geology, and combine with those branches of study which are more strictly academical, the cultivation of this new and interesting science. I am happy, therefore, in being permitted to add these expressions of my own feelings to the public respect and veneration, which have accompanied your Lordship through a long and eminently useful life; a life distinguished no less by the enlightened encouragement of learning and the liberal arts, than by the faithful discharge of the higher and more important duties which belong to your exalted station.

I have the honor to remain,

My Lord,

Your much obliged, and most devoted servant,

WILLIAM BUCKLAND.

Corpus Christi College, Oxford,
May, 1823.

CONTENTS.

PART II.

EVIDENCES OF AN INUNDATION AFFORDED BY PHENOMENA ON THE EARTH'S SURFACE.

The Description of the Cave at Kirkdale and the eleven first Plates are reprinted from the Philosophical Transactions, by permission of the President and Council of the Royal Society.

ERRATA.

Page 15, the bird not ascertained is probably a snipe.

23, *for* sheckled, *read* speckled.

82, *for* Earl Talbot, *read* C. M. Talbot, Esq.

37, *for* carcass, *read* carcase.

44, *for* hyæn aare, *read* hyæna are.

83, *for* men, *read* man.

166, *for* Uxbridge, *read* Axbridge.

220, The reported discovery of Topazes in the diluvium of Bagshot Heath
has not been verified on further examination.

Plate XXII. the figure No. 6 is omitted.

AN ACCOUNT

OF

AN ASSEMBLAGE OF FOSSIL TEETH AND BONES,

&c.

THAT portion of the present memoir which relates to the history of the cave at Kirkdale, together with a short review of its relation to other similar caves in England and Germany, has already appeared before the public in the Philosophical Transactions for 1822: the cave had been discovered in the summer of 1821, and having gone into Yorkshire for the purpose of examining it in the following December, I lost no time in laying the results of my investigation before the Royal Society of London. The encouragement they have since given me, by the award of their Copley medal, emboldens me to bring the subject again before the public in its present enlarged form; with an additional account of subsequent discoveries of several other caverns in England, and of an examination I undertook last summer of the most remarkable caves in Germany. To these I shall add a collection of facts presented by the form and structure of hills and valleys, and the accumulation on the earth's surface of diluvial loam and gravel, containing the remains of animals of the same kind with those that occur in the caverns; all tending to throw an important light on the state of

B

our planet at a period antecedent to the last great convulsion that has affected its surface, and affording one of the most complete and satisfactory chains of consistent circumstantial evidence I have ever met with in the course of my geological investigations.

As I shall have frequent occasion to make use of the word *diluvium*, it may be necessary to premise, that I apply it to those extensive and general deposits of superficial loam and gravel, which appear to have been produced by the last great convulsion that has affected our planet; and that with regard to the indications afforded by geology of such a convulsion, I entirely coincide with the views of M. Cuvier, in considering them as bearing undeniable evidence of a recent and transient inundation. On these grounds I have felt myself fully justified in applying the epithet *diluvial*, to the results of this great convulsion; of *antediluvial*, to the state of things immediately preceding it; and *postdiluvial*, or *alluvial*, to that which succeeded it, and has continued to the present time: I forbear to enter in this work into any discussion on the state or circumstances of the animal remains that occur in the solid strata that compose the surface of the earth, as it would be foreign to my immediate object, and as this subject, together with that of the days mentioned in the Mosaic account of the creation, has been already considered in my inaugural lecture published at Oxford in 1820.

In detailing the observations I am about to make on Kirkdale, I propose, first, to submit a short account of the geological position and relations of its immediate neighbourhood; to proceed, in the next place, to a description of the cavern itself; then to enter into a particular enumeration of the animal remains inhumed in it, and the very remarkable phenomena with which they are attended; to review the general inferences to which these phenomena lead; and conclude with a comparative account of analogous animal deposits, and of the evidences of diluvial action connected with them in this country and on the Continent.

Kirkdale is situated (as may be seen by reference to the annexed map, Plate I.) about 25 miles N.N.E. of the city of York, between Helmsley and Kirby Moorside, near the point at which the east base of the Hambleton hills, looking towards Scarborough, subsides into the vale of Pickering, and on the S. extremity of the mountainous district known by the name of the Eastern and the Cleveland Moorlands.

The substratum of this valley of Pickering is a mass of stratified blue clay, identical with that which at Oxford and Weymouth reposes on a similar limestone to that of Kirkdale, and containing, subordinately, beds of inflammable bituminous shale, like that of Kimeridge, in Dorsetshire. Its south boundary is formed by the Howardian hills, and by the elevated escarpment of the chalk that terminates the Wolds towards Scarborough. Its north frontier is composed of a belt of limestone, extending eastward 30 miles from the Hambleton hills, near Helmsley, to the sea at Scarborough, and varying in breadth from four to seven miles; this limestone is intersected by a succession of deep and parallel valleys (here called dales), through which the following rivers from the moorlands pass down southwards to the vale of Pickering, viz. the Rye, the Rical, the Hodge Beck, the Dove, the Seven Beck, and the Costa; their united streams fall into the Derwent above New Malton, and their only outlet is by a deep gorge, extending from near this town down to Kirkham, the stoppage of which would at once convert the whole vale of Pickering into an immense inland lake; and before the excavation of which, it is probable that such a lake existed, having its north border nearly along the edge of the belt of limestone just described, and at no great distance from the mouth of the cave at Kirkdale.

The position of the cave is at the south and lower extremity of one of these dales (that of the Hodge Beck), at the point where it falls into the vale of Pickering, at the distance of about a furlong from the church of Kirkdale, and near the brow of the left flank of the valley, close to the road. This flank slopes towards the river at an angle of 25°, and the height of the brow of the slope above the water may be about 80 feet. (See Plate II. fig. 1.)

The rock perforated by the cave is referrible to that portion of the oolite formation which, in the south of England, is known by the name of the Oxford oolite and coral rag: its organic remains are identical with those of the Heddington quarries near Oxford, but its substance is harder and more compact, and more interspersed with siliceous matter, forming irregular concretions, beds, and nodules of chert in the limestone, and sometimes entirely penetrating its coralline remains. The most compact beds of this limestone resemble the younger alpine limestone of Meillierie and Aigle in Switzerland; they alternate with and pass gradually into those of a coarser oolitic texture, and both varieties are stratified in beds from one to four feet thick. The cave is situated in one of the compact beds which lies between two others of the coarser oolitic variety; the latter vary in colour from light yellow to blue; the compact beds are of a dark grey passing to black, are extremely fetid, and full of corals and spines of the echinus cidaris. The compact portions of this oolite partake of the property common to compact limestones of all ages and formations, of being perforated by irregular holes and caverns intersecting them in all directions; the cause of these cavities has never been satisfactorily ascertained: into this question (which is one of considerable difficulty in geology) it is foreign to my present purpose to inquire, any farther than to state that they were neither produced, enlarged, nor diminished by the presence of the animals whose bones we now find in them. The half-corroded fragments of corals, shells, and spines of echini, and the irregular ledges of lime-stone and nodules of chert that project along the sides and roof of this cave, together with the small grooves and pits that cover great part of its interior, show that there was a time when its dimensions were less than they are at present, though they fail to prove by what cause it was originally produced*.

* Caves in limestone are usually more or less connected with fissures of the rock in which they exist, and the solid matter that once filled them appears in many cases to have been carried off through the fissures by the long continued and gradual perco-lation of water, removing the softer or decayed portions of the rock, either in a state of solution or mechanical suspension, so that no traces of it remain at present either in the caverns or the fissures.

I think it highly probable, from the description given by Tournefort, in his Voyage to the Levant, p. 53, and from the map and plan of it published in Sieber's Travels in Crete, 1823, Pl. 13, that the celebrated labyrinth of Crete was nothing more than a long connected series of natural caverns, such as we are now considering, a little assisted by art, and by the addition of a few corridors between the natural vaultings that compose this subterraneous wonder of classic antiquity. It is stated to occur in grey stratified limestone, which is found abundantly in Crete, and is full of caves and fissures. " Through the whole island," says Tournefort, " there are a world of caverns, especially in Mount Ida there are holes you may run your head in, bored through and through, and

The abundance of such caverns in the limestone of the vicinity of Kirkdale is evident from the fact of the engulfment of several of the rivers above enumerated in the course of their passage across it from the eastern moorlands to the vale of Pickering; and it is important to observe, that the elevation of the Kirkdale cave, above the bed of the Hodge Beck, being nearly 80 feet, excludes the possibility of our attributing the muddy sediment we shall find it to contain to any land flood or extraordinary rise of the waters of this or of any other river in the neighbourhood.

It was not till the summer of 1821, that the existence of any animal remains, or of the cavern containing them, was suspected. At this time, in continuing the operations of a large quarry along the slope just mentioned (see Plate II. fig. 1.), the workmen accidentally intersected the mouth of a long hole or cavern, closed externally with rubbish, and overgrown with grass and bushes. As this rubbish was removed before any competent person had examined

---

many very deep perpendicular abysses are also seen there;" these abysses are evidently vertical fissures connected with the caves.

    The extent to which apertures of the same kind are known to prevail in the compact limestone districts of England may be seen from a list of the principal caverns and subterraneous rivers in England, given from a note by Mr. Greenough at p. 353 of Conybeare and Phillips's Geology of England and Wales ; and from a still more detailed list given by Mr. Farey, at p. 64, and p. 292 of the 1st vol. of his Survey of Derbyshire, in which he enumerates twenty-eight remarkable caverns, and as many open fissures, locally called shake-holes, or swallow-holes, from their swallowing up the streams that cross the limestone districts of that country. The fissures descend from the surface to a depth that is very considerable, and often communicate laterally with, or enlarge themselves into, caverns.

    Similar cavities give origin very generally to the engulfment of rivers, examples of which may be seen in the limestone districts of the Mendip Hills and South Wales, of the west of Ireland, Carniola, and North America.

it, it is not certain whether it was composed of diluvial gravel and rolled pebbles, or was simply the debris that had fallen from the softer portions of the strata that lay above it; the workmen, however, who removed it, and some gentlemen who saw it, assured me, that it was composed of gravel and sand. In the interior of the cave I could not find a single rolled pebble, nor have I seen in all the collections that have been taken from it one bone, or fragment of bone, that bears the slightest mark of having been rolled by the action of water. A few bits of limestone and roundish concretions of chert that had fallen from the roof and sides, and which might be mistaken for rolled pebbles, were the only rocky fragments that I could find, with the exception of broken pieces of stalactite.

The original entrance of the cave is said to have been very small, and having been filled up as above described, there could have been no admission of external air through it to the interior of the cavern. Nearly 30 feet of its outer extremity have now been removed, and the present entrance is a hole in the perpendicular face of the quarry about three feet high and five feet broad, which it is only possible for a man to enter on his hands and knees, and which expands and contracts itself irregularly from two to seven feet in breadth and two to fourteen feet in height, diminishing however as it proceeds into the interior of the hill. The cave is about 20 feet below the incumbent field, the surface of which is nearly horizontal, and parallel to the stratification of the limestone, and to the bottom of the cave. Its main direction is E.S.E. but deviating from a straight line by several zigzags to the right and left (see Plate II. fig. 3.); its greatest length is stated by Mr. Young and Mr. Bird to be 245 feet.

In its interior it divides into several smaller passages, the extent of which has not been ascertained. In its course it is intersected by some vertical fissures, one of which is curvilinear, and again returns to the cave; another has never been traced to its termination; whilst the outer extremity of a third is probably seen in a crevice or fissure that appears on the face of the quarry, and which closes upwards before it leaves the body of the limestone. By removing the sediment and stalactite that now obstruct the smaller passages, a farther advance in them may be rendered practicable. There are but two or three places in which it is possible to stand upright, and these are where the cavern is intersected by the fissures; the latter of which continue open upwards to the height only of a few feet, when they gradually close, and terminate in the body of the limestone; they are thickly lined with stalactite, and are attended by no fault or slip of either of their sides. Both the roof and floor, for many yards from the entrance, are composed of regular horizontal strata of limestone, uninterrupted by the slightest appearance of fissure, fracture, or stony rubbish of any kind; but farther in, the roof and sides become irregularly arched, presenting a very rugged and grotesque appearance, and being studded with pendent and roundish masses of chert and stalactite; the bottom of the cavern is visible only near the entrance; and its irregularities, though apparently not great, have been filled up throughout to a nearly level surface, by the introduction of a bed of mud or loamy sediment, the history of which, and also of the stalactite, I shall presently describe *. (See Plate XVI. fig. 2.)

* For a full explanation of the terms STALACTITE, STALAGMITE, and BRECCIA, which I shall frequently make use of, I beg to refer my readers to Dr. Kidd's Outlines of

The fact already mentioned of the engulfment of the Rical Beck, and other adjacent rivers, as they cross the limestone, showing it to abound with many similar cavities to those at Kirkdale, renders it likely that other deposits of bones may hereafter be discovered in this same neighbourhood; but as the mouths of these caverns are filled up and buried under diluvial loam and gravel, or post-diluvian detritus, and overgrown with grass, nothing but their casual inter-section by some artificial operations will lead to a knowledge of their existence; and in this circumstance, we also see a reason why so few caverns of this kind have hitherto been discovered, although it is probable that they are very numerous.

In all these cases, the bones found in caverns are never mineral-ised, but simply in the state of grave bones more or less decayed or incrusted by stalagmite; and have no farther connexion with the rocks themselves, than that arising from the accident of having been

---

Mineralogy: the following extract will suffice for my present purpose. Water, in pene-trating through limestone strata, often becomes impregnated with particles of the cal-careous carbonate of which the limestone is composed, and which on exposure to air it again deposits either in the form of pendulous masses that hang like icicles from the roof, or of stony concretions adhering to the sides of cavities, into which the water thus impregnated finds admission: to such deposits the term STALACTITE is applied.

If the percolation of water containing calcareous particles is too rapid to allow time for the formation of a stalactite, the earthy matter is deposited from it after it has fallen from the roof upon the floor of the cavern; and in this case the deposition is called STALAGMITE: the substance deposited is the same as in the case of stalactite. Stalagmites are commonly, at least in the early stages of their formation, of a mamillary shape; by gradual accumulation they become conical; and at length form pillars, by the continual addition of their materials, till they meet, and become united with the stalactite that depends from the roof immediately above.

The term BRECCIA is applied to broken fragments of stone or bone reunited into a solid mass by a stony cement.

c

lodged in their cavities, at periods long subsequent to the formation and consolidation of the strata in which these cavities occur.

On entering the cave at Kirkdale (see Plate II. fig. 2.), the first thing observed was a sediment of soft mud or loam, covering entirely its whole bottom to the average depth of about a foot, and concealing the subjacent rock, or actual floor of the cavern. Not a particle of mud was found attached either to the sides or roof; nor was there a trace of it adhering to the sides or upper portions of the transverse fissures, or any thing to suggest the idea that it entered through them. The surface of this sediment when the cave was first opened was nearly smooth and level, except in those parts where its regularity had been broken by the accumulation of stalagmite above it, or ruffled by the dripping of water: its substance is an argillaceous and slightly micaceous loam, composed of such minute particles as would easily be suspended in muddy water, and mixed with much calcareous matter, that seems to have been derived in part from the dripping of the roof, and in part from comminuted bones. At about 100 feet within the cave's mouth the sediment became more coarse and sandy, and partially covered with an incrustation of black manganese ore.

Above this mud, on advancing some way into the cave, the roof and sides were found to be partially studded and cased over with a coating of stalactite, which was most abundant in those parts where the transverse fissures occur, but in small quantity where the rock is compact and devoid of fissures. Thus far it resembled the stalactite of ordinary caverns; but on tracing it downwards to the surface of the mud, it was there found to turn off at right angles from the sides of the cave, and form above the mud a plate or crust, shooting across like ice on the surface of water, or cream on a pan of milk. (See Plate II. fig. 2.) The thickness and quantity of this crust varied with that found on the roof and sides, being most abundant, and covering the mud entirely where there was much stalactite on the sides, and more scanty in those places where the roof or sides presented but little: in many parts it was totally wanting both on the roof and surface of the mud and of the subjacent floor. Great portion of this crust had been destroyed in digging up the mud to extract the bones before my arrival; it still remained, however, projecting partially in some few places along the sides; and in one or two, where it was very thick, it formed, when I visited the cave, a continuous bridge over the mud entirely across from one side to the other. In the outer portion of the cave, there was originally a mass of this kind which had been accumulated so high as to obstruct the passage, so that a man could not enter till it had been dug away.

These horizontal incrustations have been formed by the water which, trickling down the sides, was forced to ooze off laterally as soon as it came into contact with the mud; in other parts, where it fell in drops from the roof, stalagmitic accumulations have been raised on its surface, some of which are very large and flat, resembling a cake of bees wax, but more commonly they are of the size and shape of a cow's pap, a name which the workmen have applied to them. There is no alternation of mud with any repeated beds of stalactite, but simply a partial deposit of the latter on the floor beneath it; and it was chiefly in the lower part of the earthy sediment, and in the stalagmitic matter beneath it, that the animal remains were found:

c 2

there was nowhere any black earth or admixture of animal matter, except an infinity of extremely minute particles of undecomposed bone. In the whole extent of the cave, only a very few large bones have been discovered that are tolerably perfect; most of them are broken into small angular fragments and chips, the greater part of which lay separately in the mud, whilst others were wholly or partially invested with stalagmite; and others again mixed with masses of still smaller fragments and cemented by stalagmite, so as to form an osseous breccia. In some few places where the mud was shallow, and the heaps of teeth and bones considerable, parts of the latter were elevated some inches above the surface of the mud and its stalagmitic crust; and the upper ends of the bones thus projecting like the legs of pigeons through a pie-crust into the void space above, have become thinly covered with stalagmitic drippings, whilst their lower extremities have no such incrustation, and have simply the mud adhering to them in which they have been imbedded; an horizontal crust of stalagmite, about an inch thick, crosses the middle of these bones, and retains them firmly in the position they occupied at the bottom of the cave. A large flat plate of stalagmite, corresponding in all respects with the above description, and containing three long bones fixed so as to form almost a right angle with the plane of the stalagmite, is in the collection of the Rev. Mr. Smith, of Kirby Moorside. The same gentleman has also, among many other valuable specimens, a fragment of the thigh bone of an elephant, which is the largest I have seen from this cave.

The effect of the loam and stalagmite in preserving the bones from decomposition, by protecting them from all access of atmospheric air, has been very remarkable; some that had lain uncovered in the cave for a long time before the introduction of the loam were in various stages of decomposition; but even in these the further progress of decay appears to have been arrested as soon as they became covered with it; and in the greater number, little or no destruction of their form, and scarcely any of their substance, has taken place. I have found, on immersing fragments of these bones in an acid till the phosphate and carbonate of lime were removed, that nearly the whole of their original gelatine has been preserved. Analogous cases of animal remains preserved from decay by the protection of similar diluvial mud occur on the coast of Essex, near Walton, and at Lawford, near Rugby, in Warwickshire; here the bones of the same species of elephant, rhinoceros, and other diluvial animals occur in a state of freshness and perfection even exceeding that of those in the cave at Kirkdale; and from a similar cause, viz. their having been guarded from the access of atmospheric air, or the percolation of water, by the argillaceous matrix in which they have been imbedded: whilst other bones that have lain the same length of time in diluvial sand, or gravel, and been subject to the constant percolation of water, have lost their compactness and strength, and great part of their gelatine, and are often ready to fall to pieces on the slightest touch; and this where the beds of clay and gravel in question alternate in the same quarry, as at Lawford.

The workmen on first discovering the bones at Kirkdale, supposed them to have belonged to cattle that died by a murrain in this district a few years ago, and they were for some time neglected, and thrown on the roads with the common limestone; they were at length noticed by Mr. Harrison, a medical gentleman of Kirby Moorside, and have since been collected and dispersed amongst so many individuals, that it is probable nearly all the specimens will in a few years be lost, with the exception of such as may be deposited in public collections. By the kindness and liberality of the Bishop of Oxford (to whom I am also indebted for my first information of the existence of this cave) and of C. Duncombe, Esq. and Lady Charlotte Duncombe, of Duncombe Park, a nearly complete series of the teeth discovered in it has been presented to the Museum at Oxford; whilst a still better collection, both of teeth and bones, is in the possession of J. Gibson, Esq. of Stratford in Essex, to whose exertions we owe the preservation of many valuable specimens, and who has presented a series of them to several public collections in London *. W. Salmond, Esq. also, of York, has been engaged with much zeal and activity in measuring and exploring new branches of the cave, and making large collections of the teeth and bones, from which he has sent specimens to the Royal Institution of London and to M. Cuvier. He has recently deposited the bulk of his collection at the newly-established Philosophical Society at York. I am indebted to him for the annexed ground plan of the cave, and its ramifications,

* The British Museum, the Royal College of Surgeons, and the Geological Society have all been enriched by the liberality of Mr. Gibson. The Geological Society possesses also a magnificent collection of the remains of elephant, rhinoceros, ox, elk, and other anteluvian animals found in the diluvian gravel beds of various parts of England, together with some fine specimens of bones from the caverns of Germany: their collection also of the organic remains found in the secondary strata of England, and of specimens of the strata themselves, is arranged in a manner which affords to the members of that society the most ready access to a knowledge of the physical changes which the country we inhabit has undergone, and of general geology.

(Plate II. fig. 3.)   Drawings by Mr. Clift, of some of the most perfect of Mr. Gibson's specimens, have been sent to M. Cuvier, for the new edition of his work on fossil animals; copies of these have been made for me by the kindness of Miss Morland, and appear in the annexed plates, with many other drawings, for which I am indebted to the pencil of Miss Duncombe; and the Rev. George Young, and Mr. Bird of Whitby, in their History of the Geology of the Coast of Yorkshire, have given engravings of some other teeth and a few bones in their possession.

It appears that the teeth and bones which have as yet been discovered in the cave at Kirkdale are referable to the following 23 species of animals.

6 Carnivora.—Hyæna, Tiger, Bear, Wolf, Fox, Weasel. (See Plates III. IV. V. VI. and XIII.)

4 Pachydermata.—Elephant, Rhinoceros, Hippopotamus, and Horse.   (See Plates VII. X. and XIII.)

4 Ruminantia.—Ox, and three species of Deer.   (See Plates VIII. IX. and X.)

4 Rodentia.—Hare, Rabbit, Water-rat, and Mouse.   (See Plates X. XI. and XIII.)

5 Birds.—Raven, Pigeon, Lark, a small species of Duck, resembling the anas sponsor, or summer duck, and a Bird not ascertained, being about the size of a thrush.   (See Plate XI. fig. 19 to 29, and Plate XIII. fig. 11, 12.)

The bottom of the cave, on first removing the mud, was found to be strewed all over like a dog-kennel, from one end to the other, with hundreds of teeth and bones, or rather broken and splintered fragments of bones, of all the animals above enumerated; they were found in greatest quantity near its mouth, simply because its area in this part was most capacious; those of the larger animals, elephant, rhinoceros, &c. were found co-extensively with all the rest, even in the inmost and smallest recesses. (See Plate II. fig. 3.*) Scarcely a single bone has escaped fracture, with the exception of the astragalus, and other hard and solid bones of the tarsus and carpus joints, and those of the feet. (See Plate X. fig. 1 to 5, and fig. 7 to 10; and Plate V. fig. 5 to 12.) On some of the bones, marks may be traced, which, on applying one to the other, appear exactly to fit the form of the canine teeth of the hyæna that occur in the cave. The hyæna's bones have been broken, and apparently gnawed equally with those of the other animals. Heaps of small splinters, and highly comminuted, yet angular fragments of bone, mixed with teeth of all the varieties of animals above enumerated, lay in the bottom of the den, occasionally adhering together by stalagmite, and forming, as has been before mentioned, an osseous breccia. Many insulated fragments also are wholly or partially enveloped with stalagmite, both externally and internally. Not one skull is to be found entire; and it is so rare to find a large bone of any kind that has not been more or less broken, that there is no hope of obtaining materials for the construction of a single limb, and still less of an entire skeleton. The jaw bones also, even of the hyænas, are broken to pieces like the rest; and in the case of all the animals, the number of teeth and of solid bones of the tarsus and carpus is more than twenty times as great as could have been supplied by the individuals whose other bones we find mixed with them.

Fragments of jaw bones are by no means common; the greatest number I saw belong to the deer, hyæna, and water-rat, and retain their teeth; in all the jaws both teeth and bone are in an equal state of high preservation, and show that their fracture has been the effect of violence, and not of natural decay. I have seen but ten fragments of deers' jaws, and about forty of hyænas, and as many of rats. (See Plate III. fig. 3, 4, 5, and Plate IV. fig. 2, 3). The ordinary fate of the jaw bones, as of all the rest, appears to have been to be broken to pieces.

The greatest number of teeth are those of hyænas, and the ruminantia. Mr. Gibson alone collected more than 300 canine teeth of the hyæna, which at the least must have belonged to 75 individuals, and adding to these the canine teeth I have seen in other collections, I cannot calculate the total number of hyænas of which there is evidence, at less than 200 or 300. I have already stated, that many of these animals had died before the first set, or milk teeth, had been shed; these teeth are represented in Plate VI. fig. 15 to 27: the state of their fangs shows that they had not fallen out by absorption. The only remains that have been found of the tiger species (see Plate VI. fig. 5, 6, 7) are two large canine teeth, each four inches in length, and a few molar teeth, one of which is in my possession; these exceed in size that of the largest lion or Bengal tiger. There is one tusk only of a bear (see Plate VI. fig. 1), which exactly resembles those of the extinct ursus spelæus of the caves of Germany, the size of which M. Cuvier says must have equalled that of a large

horse. Of the fox there are many teeth (see Plate VI. fig. 8 to 14).
Of the wolf I do not recollect that I have seen more than one large
molar tooth (see Plate XIII. fig. 5, 6); the smaller molars of the wolf
however are very like some of the first set of the young hyæna. A
few jaws and teeth have also been found belonging to the weasel.
(Plate VI. fig. 28, 29). Teeth of the larger pachydermatous animals
are not abundant. I have information of about ten elephants' teeth,
but of no tusk; most of these teeth are broken, and as very few of
them exceed three inches in their longest diameter, they must have
belonged to extremely young animals. (See Plate VII. fig. 1 and 2).
I have seen but six molar teeth of the hippopotamus, and a few
fragments of its canine and incisor teeth, the best of which are in the
possession of Mr. Thorpe, of York. (See Plate VII. fig. 8, 9, 10,
and Plate XIII. fig. 7). Teeth of the rhinoceros are not so rare: I
have seen at least 50, some of them very large, and apparently from
aged animals. (See Plate VII. fig. 3, 4, 5, 6). I have heard of only
two or three teeth belonging to the horse. Of the teeth of deer
there are at least three species (see Plate VIII. fig. 9, 11, 13), the
smallest being very nearly of the size and form of those of a fallow
deer, the largest agreeing in size, but differing in form from those of
the modern elk; and a third being of an intermediate size, and
approaching that of a large stag or red deer. I have not ascertained
how many species there are of ox, but apparently there are two.
But the teeth which occur perhaps in greatest abundance are those
of the water-rat (see Plate XI. fig. 1 to 6, and 11 to 18); for in almost
every specimen I have collected or seen of the osseous breccia, there
are teeth or broken fragments of the bones of this little animal mixed

with and adhering to the fragments of all the larger bones.   These
rats may be supposed to have abounded on the edge of the lake,
which I have shown probably existed at that time in this neighbourhood: there are also the jaw of a hare, and a few teeth and bones of
rabbits and mice.   (Plate X. fig. 14, 15, 16, 17, 18, Plate XI. fig.
7, 8, 9, 10, and Plate XIII. fig. 8).

Besides the teeth and bones already described, the cave contained
also fragments of horns of at least two species of deer. (See Plate IX.
fig. 3, 4, and 5).   One of these resembles the horn of the common
stag or red deer, the circumference of the base measuring $9\frac{3}{4}$ inches,
which is about the size of our largest stag.   A second (fig. 4) measures $7\frac{3}{4}$ inches at the same part, and both have two antlers, that rise
very near the base.   In a smaller species the lowest antler is $3\frac{1}{2}$ inches
above the base, the circumference of which is 8 inches.   (See fig. 5).
No horns are found entire, but fragments only, and these apparently
gnawed to pieces like the bones: their lower extremity nearest the
head is that which has generally escaped destruction: and it is a
curious fact, that this portion of all the horns I have seen from the
cave shows, by the rounded state of the base, that they had fallen
off by absorption or necrosis, and been shed from the head on which
they grew, and not broken off by violence.

It must already appear probable, from the facts above described,
particularly from the comminuted state and apparently gnawed condition of the bones, that the cave at Kirkdale was, during a long
succession of years, inhabited as a den by hyænas, and that they
dragged into its recesses the other animal bodies whose remains are
found mixed indiscriminately with their own: this conjecture is ren-

dered almost certain by the discovery I made, of many small balls
of the solid calcareous excrement of an animal that had fed on bones,
resembling the substance known in the old Materia Medica by the
name of album græcum (see Plate X. fig. 6): its external form is that
of a sphere, irregularly compressed, as in the fæces of sheep, and
varying from half an inch to an inch and half in diameter; its colour
is yellowish white, its fracture is usually earthy and compact, resembling steatite, and sometimes granular; when compact, it is
interspersed with small cellular cavities, and in some of the balls
there are undigested minute fragments of the enamel of teeth.   It
was at first sight recognised by the keeper of the Menagerie at
Exeter Change, as resembling, both in form and appearance, the
fæces of the spotted or Cape hyæna, which he stated to be greedy of
bones beyond all other beasts under his care.   This information I
owe to Dr. Wollaston, who has also made an analysis of the substance under discussion, and finds it to be composed of the ingredients that might be expected in fæcal matter derived from bones,
viz. phosphate of lime, carbonate of lime, and a very small proportion
of the triple phosphate of ammonia and magnesia; it retains no
animal matter, and its originally earthy nature and affinity to bone
will account for its perfect state of preservation*.

I do not know what more conclusive evidence than this can be
added to the facts already enumerated, to show that the hyænas

---

* I have one ball of this substance that is in great part invested with a thin circular
case or crust of stalagmite. This must have been formed round it whilst it lay loose and
exposed to the dripping of water on the bottom of the cave, before the introduction of
the mud.

inhabited this cave, and were the agents by which the teeth and bones of the other animals were there collected; it may be useful therefore to consider, in this part of our inquiry, what are the habits of modern hyænas, and how far they illustrate the case before us.

The modern hyæna (of which there are only three known species, all of them smaller and different from the fossil one) is an inhabitant exclusively of hot climates; the most savage, or striped species, abounds in Abyssinia, Nubia, and the adjacent parts of Africa and Asia. The less ferocious, or spotted one, inhabits the Cape of Good Hope, and lives principally on carrion. He is seldom seen by day, but prowls by night, and clears the plains of the carcasses, and even skeletons, which the vultures have picked clean, in preference to attacking any living creature. In the structure of its bones this animal approaches more nearly than the striped hyæna to the fossil species: to these M. Cuvier adds a third, the red hyæna, which is very rare.

The structure of these animals places them in an intermediate class between the cat and dog tribes; not feeding, like the former, almost exclusively on living prey, but like the latter, being greedy of putrid flesh and bones*: their love of putrid flesh induces them to follow armies, and dig up human bodies from the grave. They inhabit holes in the earth, and chasms of rocks; are fierce, and of obstinate courage, attacking stronger quadrupeds than themselves, and even repelling lions. Johnson says of them, in his Field Sports,

---

* It is quite impossible to mistake the jaw of any species of hyæna for that of the wolf or tiger kind; the latter having three molar teeth only in the lower jaw, and the former seven; whilst all the hyæna tribe have four. (See Plate IV. fig. 1, 2, 3).

that " they feed on small animals and carrion, and often come in for the prey left by tigers and leopards after their appetites have been satiated: they are great enemies of dogs, and kill numbers of them. They make no earths of their own, but lie under rocks, or resort to the earths of wolves, as foxes do to those of badgers; and it is not uncommon to find wolves and hyænas in the same bed of earths." Their habit of digging human bodies from the grave, and dragging them to their den, and of accumulating around it the bones of all kinds of animals, is thus described by Busbequius, where he is speaking of the Turkish mode of burial in Anatolia, and their custom of laying large stones upon their graves to protect them from the hyænas. " Hyæna regionibus iis satis frequens; sepulchra suffodit, extrahitque cadavera, portatque ad suam speluncam; juxta quam videre est ingentem cumulum ossium humanorum ' veterinariorum' * et reliquorum omne genus animalium." (Busbeq. Epist. 1 Leg. Turc.†) Brown, also, in his Travels to Darfur, describes the hyænas' manner of taking off their prey in the following words:—" they come in herds of six, eight, and often more, into the villages at night, and carry off with them whatever they are able to master; they will kill dogs and asses even within the enclosure of houses, and fail not to assemble wherever a dead camel or other animal is thrown, which, acting in concert, they sometimes drag to a prodigious distance."

* Veterinam bestiam jumentum Cato appellavit a vehendo : (quasi veheterinus vel veterinus.) Pomp. Fest.

† This evidence is the more valuable, from the accuracy and delight with which it appears, from his own testimony, that Busbequius used to watch the habits of wild animals, which he kept for this purpose in his menagerie at Constantinople, where he resided many years as ambassador from the Emperor of Germany.

Sparman and Pennant mention that a single hyæna has been known to carry off a living man or woman in the vicinity of the Cape *.

The strength of the hyæna's jaw is such, that in attacking a dog, he begins by biting off his leg at a single snap. The capacity of his teeth, for such an operation, is sufficiently obvious from simple inspection; and, consistent with this strength of teeth and jaw, is the

---

* It appears from the discussions of the learned Bochart, in his Hierozoicon, on the hyæna, that the peculiar habits of this animal had attracted the attention of the earliest naturalists, more especially his savage voracity, and practice of digging human bodies from their graves for the purpose of devouring them. He quotes the following passages: Aristotelis Hist. lib. viii. cap. 5. " Τυμβωρυχεῖ δὲ ἐφιέμενον τῆς σαρκοφαγίας τῶν ἀνθρώπων."— Plinius, lib. viii. cap. 30. " Ab uno animali sepulchra erui (traduntur) inquisitione corporum."—Solinus, " Eadem hyæna inquisitione corporum sepultorum busta eruit."— Hieronymus in Esaiam, capite lxv. " Semper cadavera persequitur et vivit succo et sanie corporum mortuorum."—Et in Ieremiam, capite xiii. " Vivit cadaveribus mortuorum, et de sepulchris solet effodere corpora."

Bochart shows also that certain parts of the body of this animal, particularly the atlas or first vertebra of the neck, which they called the " nodus," were used by the ancient enchanters in the ceremonies of their magical incantations.

> " Huc quicquid fœtu genuit natura sinistro
> Miscetur: non spuma canum, quibus unda timori est;
> Viscera non lyncis, non diræ nodus hyænæ
> Defuit."

Lucanus, Lib. VI. v. 673.

And contends that the same animal is also alluded to in the Old Testament, in 1 Samuel, ch. 13. v. 18, and Jeremiah, ch. 12. v. 9.

In the former of these passages he is of opinion with Aquila, that the " Valley of Zeboim" ought to have been translated the " Valley of Hyænas;" and in the latter he thinks with the Septuagint, that the words which in our version are rendered " sheckled bird," should have been " ravenous spotted beast," i. e. hyæna. The Septuagint have it, " Μὴ σπήλαιον ὑαίνης ἢ κληρονομία μυ ἐμοί." Mr. Parkhurst, also, and Scheuzer are for establishing the hyæna in this passage.

The proverbial enmity supposed to subsist between this animal and the dog is also mentioned by Oppian, Pliny, and Ælian, and alluded to in Ecclesiasticus, ch. xiii. 18. "Τίς εἰρήνη ὑαίνη πρὸς κύνα;"—" What agreement is there between the hyæna and a dog?"

state of the muscles of his neck, being so full and strong, that in early times this animal was fabled to have but one cervical vertebra. They live by day in dens, and seek their prey by night, having large prominent eyes, adapted, like those of the rat and mouse, for seeing in the dark.   To animals of such a class, our cave at Kirkdale would afford a most convenient habitation; and the circumstances we find developed in it are entirely consistent with the habits above enumerated.

It appears, from the researches of M. Cuvier, that the fossil hyæna was nearly one-third larger than the largest of the modern species, that is, the striped or Abyssinian; but, in the structure of its teeth, more nearly resembled that of the Cape animal.   (See Plate III. fig. 1, 2, 3, 4, and Plate IV. fig. 1, 2, 3.)   Its muzzle also was shorter and stronger than in either of them, and consequently its bite more powerful.   The length of the largest modern hyæna noticed is five feet nine inches.

The fossil species has been found on the Continent in situations of two kinds, both of them consistent with the circumstances under which it occurs in Yorkshire, and, on comparing the jaws and teeth of the latter with those of the former engraved in M. Cuvier's Recherches sur les Ossemens fossiles, I find them to be absolutely identical.   The two situations are caverns and diluvian gravel.

1. In Franconia the bones of hyæna have been found mixed with those of an enormous number of bears and of tigers in the caves near Muggendorf.

2. In the Hartz Forest similar bones of hyæna, bears, and tigers have been found together in the caves of Scharzfeld and Bauman's Hole.

3. At Sundwick, in Westphalia, Mr. A. L. Sack, of Bonn, has within the last two years discovered the remains of hyænæ in the same cave with the bones of the ursus spelæus, ursus arctoideus, and glutton; and accompanied by the molar teeth and foot bones of rhinoceros, and the horns, jaws, and other bones of two species of large deer; the bones of the deer and rhinoceros he describes as having their softer parts broken off, and bearing distinct marks of the teeth of the wild beasts by which they have been gnawed.

4. In France, at Fouvent, near Gray, in the department of Doubes, bones of hyænæ were found mixed with the teeth and tusks of elephants, and the bones of the rhinoceros and horse, in a cavity of limestone rock, which, like that at Kirkdale, was discovered by the accidental digging away of the rock in a garden.

5. In Saxony, on the S. W. of Leipsig, Baron Schlotheim has discovered at Köstritz, in the valley of the Elster, the bones of hyænæ mixed with those of rhinoceros, horse, ox, stag, bear, and extinct tiger, in the fissures and cavities both of the limestone and gypsum rocks which occur in that district. The bones are buried in, and mixed up with a mass of diluvial loam or clay, containing also pebbles of limestone and granite.

6. In Wirtemberg at Canstadt, in the valley of the Necker, A. D. 1700, hyænas' bones were found mixed with those of the elephant, rhinoceros, horse, ox, stag, hare, and small carnivorous animals, and with rolled pebbles, in a mass of

E

yellowish clay and sand, which was partially agglutinated into a hard breccia: they lay without any order, or relative proportion to each other, were for the most part broken, and some of them rolled.

7. In Bavaria between Kahldorf and Reiterbuck, on the surface of the hills that bound the valley of Eichstadt. These were buried in a bed of sand, and mixed with the bones of elephants and stags.

8. On the west base of the Hartz Forest, Blumenbach has described, in his Specimen, 2. Archæologiæ Telluris, a mass of bones belonging to the hyæna, elephant, and rhinoceros, discovered in 1808 between Osterode and Dorste, and imbedded in diluvial mud.

9. In Italy, M. Pentland has found, in the Museum at Florence, the head and lower jaw of an hyæna, and the remains of bear, tiger, elephant, rhinoceros, and hippopotamus, from the sand and loam beds of the Val d'Arno.

The three first of these cases appear to have been dens like the cave at Kirkdale; the 4th and 5th have possibly received their bones in the manner I shall hereafter point out when speaking of the caves at Plymouth; the four last are deposits of diluvial detritus, like the surface gravel beds of England.

The bones of the hyæna, however, had not been discovered in the diluvial detritus of this country till the spring of last year, 1822, when Mr. Andrew Bloxham, by mere accident, brought me some bones from the clay in which they so often find the remains of elephant and rhinoceros at Lawford, near Rugby, that I might inform him what they were: the instant I saw them I was enchanted to find the entire under jaw and entire radius and ulna of a very old and large hyæna, supplying the only link that was deficient to complete the evidence I wanted to establish the hyæna's den at Kirkdale. These bones are in the highest possible state of preservation; the jaw is quite entire, and from an animal so old that it had lost half its teeth, and the remainder are ground almost to the stumps. (See Plate XII.) The bones of the arm also (see Plate XIII. fig. 1, 2), are equally perfect with the jaw. There are not the slightest marks of fracture on any of them like those on the bones at Kirkdale; and this is consistent with the different circumstances of this individual from those in the cave; the hyæna at Lawford appears from its position in the diluvial clay to have been one that perished by the inundation that extirpated the race, as well as the elephant, rhinoceros, and other tribes that lie buried with it; and, consequently, as it could have had no survivors to devour its bones, we should on this hypothesis expect to find them entire, as they are actually found in the specimens before us. (See Plate XII. and Plate XIII. fig. 1, 2.) With them were found some entire small bones of the foot apparently of the same individual hyæna, and also the humerus of a bird in size and shape nearly resembling that of a goose (see Plate XIII. fig. 9, 10), and in the same state of high preservation with the hyæna and rhinoceros bones amidst which it lay. This is the first example within my knowledge of the bones of birds being noticed in the diluvium of England.

It was stated, when speaking of the den at Kirkdale, that the bones of the hyænas are as much broken to pieces as those of the animals that formed their prey; and hence we must infer, that the carcases even of the hyænas themselves were eaten up by their survivors.   Whether it be the habit of modern hyænas to devour those of their own species that die in the course of nature; or, under the pressure of extreme hunger, to kill and eat the weaker of them, is a point on which it is not easy to obtain positive evidence.   Mr. Brown however asserts, in his journey to Darfur, " that it is related of the hyænas, that upon one of them being wounded, his companions instantly tear him to pieces and devour him."   It seems therefore in the highest degree probable, that the mangled relics of hundreds of hyænas that lie indiscriminately scattered and equally broken with the bones of other animals in the cave of Kirkdale were reduced to this state by the agency of the surviving individuals of their own species.

It has not only been stated as above, that modern hyænas devour their own species, but still further, that when in captivity they eat up parts of their own bodies.   The keeper of Mr. Wombwell's collection told me in December last, that he had an hyæna some years ago which ate off his own fore paws; and his Royal Highness the Prince of Denmark informed me last summer, that the old hyæna in the Jardin du Roi at Paris has eaten off his own hind feet.   I have since requested my friend Mr. Underwood, who is resident in Paris, and is an accurate observer of nature, to procure me further particulars of this circumstance, which I subjoin in his own words.

" The present Cape hyæna," says he, " about ten years ago, in the month of September, began to nibble and suck his hinder paws, which nearly destroyed them in two months, at which time he left off: at the same period of the following year he began again, and continued for about the same space of time, by which the metatarsal and tarsal bones of both feet, and about half the tibia and fibula of the right leg, were eaten. Since that time he has not attacked any other part of his body. He now walks on three legs, but with great difficulty." The fact seems to be, that many animals, particularly the monkey tribe, when in confinement, are subject to a sort of itching, which induces them to nibble their extremities; and in the case of monkeys, especially their tails, and that they rarely cease until mortification and death ensue.

A large proportion of the hyænas' teeth bear marks of extreme old age, some being abraded to the very sockets by continual gnawing, and the majority having lost the upper portion of their coronary part, and having fangs extremely large: these probably died in the den from mere old age: and if we compare the lacerated condition of the bones that accompany them, with the state of the teeth thus worn down to the very stumps, notwithstanding their prodigious strength, we find in the latter the obvious instruments by which the former were thus comminuted. A great number of other teeth appear to have belonged to young hyænas (see Plate VI. figs. 15 to 27), being of the first set or milk teeth; in many others of the second set the fangs are not developed, and the points and edges of the crown not the least worn. I have a fragment of the lower jaw, in which the second set of teeth had not been protruded, but were in the act of forming below the first. (See Plate V. fig. 3, 4.) Mr. Salmond has an entire one. (See Plate XXIII. fig. 7.) Archdeacon Wrangham has a similar fragment of an upper jaw (see Plate XIII. fig. 3, 4), where the second set of teeth are in the act of forcing out one of the first set, which is identical with those engraved in Plate VI. No. 22, 23, 24, 25; and I have seen at least a dozen smaller fragments of jaws, which are nearly in the same state; other teeth of the second set are found in various stages of advancement towards maturity, and the number of all these young teeth is much too great for us to attribute them to animals that may have died in early life from accident or disease. It seems more probable, and the idea is confirmed by the above statement of Mr. Brown, and by the fact of all the hyænas' bones in the den being gnawed and broken to pieces equally with the rest, that they were occasionally killed and devoured by the stronger individuals of their own species, and that both young and old were always eaten up after natural death.

But, besides the evidence their teeth afford to show that the animals died at various periods of life, they present other appearances (and so likewise do the bones) of having passed through different stages and gradations of decomposition, according to the different length of time they had lain exposed in the bottom of the den, before the muddy sediment entered, which, since its introduction, has preserved them from farther decay. This observation applies equally to the remains of all the animals. I have some portions of bone and teeth that are so much decomposed as to be ready to fall to pieces on the slightest touch: these had probably lain a long time unprotected in the bottom of the den; others still older may have entirely perished; but the majority both of teeth and fragments of bone are in a state of the highest preservation; and many thousands have been collected and carried away since the cave was discovered. The degree of decay is always equal in the teeth and portions of jaw bones to which they are attached.

In many of the most highly preserved specimens of teeth and

bones there is a curious circumstance, which, before I visited Kirkdale, had convinced me of the existence of the den, viz. a partial
polish and wearing away to a considerable depth of one side only;
many straight fragments of the larger bones have one entire side, or
the fractured edges of one side, rubbed down and worn completely
smooth, whilst the opposite side and ends of the same bone are sharp
and untouched, in the same manner as the upper portions of pitch-
ing stones in the street become rounded and polished, whilst their
lower parts retain the exact form and angles which they possessed
when first laid down.   This can only be explained by referring the
partial destruction of the solid bone to friction from the continual
treading of the hyænas, and rubbing of their skin on the side that
lay uppermost in the bottom of the den.   In many of the smaller
and curved bones, also, particularly in those of the lower jaw,
(see Plate V. fig. 1 and 3) the convex surface only has been
uniformly worn down and polished, whilst the ends and concave
surface have suffered no kind of change or destruction, (Plate V.
fig. 2 and 4): and this also admits of a similar explanation; for
the curvature of the bone would allow it to rest steady under constant treading only in this position: as long as the concave surface
was uppermost, pressure on either extremity would cause it to tilt
over, and throw the convex side upwards; and this done, the next
pressure would cause its two extremities to sink into any soft substance that lay beneath, and give it a steady and fixed position.
Such seems to have been the process by which the curved fragments
I allude to have not only received a partial polish on the convex
side only, but have been submitted to so much friction, that in

several instances more than one-fourth of the entire thickness of the bone, and a proportionate quantity of the outer side of the fangs and body of the teeth, have been entirely worn away. (See Plate V. fig. 1.) I can imagine no other means than the repeated touch of the living hyænas' feet and skin, by which this partial wearing away and polish can have been produced *: for the process of rolling by water would have made pebbles of them, or at least would have broken off the edges of the teeth and delicate points of the fractured extremities of the bone, which still remain untouched and sharp.

I have already stated, that the greatest number of teeth (those of the hyæna excepted) belong to the ruminating animals ; from which it is to be inferred that they formed the ordinary prey of the hyænas. I have also to add, that very few of the teeth of these animals bear marks of age ; they seem to have perished by a violent death in the vigour of life. With respect to the horns of deer that appear to have fallen off by necrosis, it is probable that the hyænas found them thus shed, and dragged them home for the purpose of gnawing them in their den ; and to animals so fond of bones, the spongy interior of horns of this kind would not be unacceptable. I found a fragment of stag's horn in so small a recess of the cave, that it never could have been introduced, unless singly, and after separation from the head;

* I have been informed by an officer in India, that passing by a tiger's den in the absence of the tiger, he examined the interior, and found in the middle of it a large portion of stone, on which the tiger reposed, to be worn smooth and polished by the friction of his body. The same thing may be seen on marble steps and altars, and even metallic statues in places of worship that are favourite objects of pilgrimage: they are often deeply worn and polished by the knees, and even lips of pilgrims, to a degree that, without experience of the fact, we could scarcely have anticipated

and near it was the molar tooth of an elephant. I have seen no remains of the horns of oxen, and perhaps there are none; for the bony portion of their interior, being of a porous spongy nature, would probably have been eaten by the hyænas, whilst the outer case, being of a similar composition to hair and hoofs, would not long have escaped total decomposition. For the same reason the horn of the rhinoceros, being merely a mass of compacted hair-like fibres, has never been found fossil in gravel beds with the bones of that animal, except in Siberia, where it has been frozen up in ice, nor does it occur in the cave at Kirkdale. I have been told that sheeps' horns laid on land for manure will be consumed in ten or a dozen years; the calcareous matter of bone, being nearly allied to limestone, is the only portion of animal bodies that occurs in a fossil state, unless when preserved, like the Siberian elephant, of the same extinct species with that of Kirkdale, by being frozen in ice, or buried in peat.

The extreme abundance of the teeth of water rats has also been alluded to; and though the idea of hyænas eating rats may appear ridiculous, it is consistent with the omnivorous appetite of modern hyænas, and with the fact, quoted from Johnson, that they feed on small animals, as well as carrion and bones; nor is the disproportion in size of the animal to that of its prey greater than that of wolves and foxes, which are supposed by Captain Parry to feed chiefly on mice, during the long winters of Melville Island. Hearne, in his Journey to the Northern Ocean, mentions the fact " of a hill, called Grizzle Bear Hill, being deeply furrowed and turned over like ploughed land by bears in search of ground squirrels, and perhaps mice, which constitute a favourite part of their food." If bears eat

F

mice, why should not hyænas eat rats? Our largest dogs eat rats and mice; jackalls occasionally prey on mice, and dogs and foxes will eat frogs. It is probable, therefore, that neither the size nor aquatic habit of the water rat would secure it from the hyænas. They might occasionally also have eaten mice, weasels, rabbits, foxes, and birds; and in masticating the bodies of these small animals with their coarse conical teeth, many bones and fragments of bone would be pressed outwards through their lips, and fall neglected to the ground *.

The occurrence of birds' bones may be explained by the probability of the hyænas finding the birds dead, and taking them home, as usual, to eat in their den: and the fact, that four of the only six bones of birds I have seen from Kirkdale are those of the ulna, may have arisen from the position of the quill feathers on it, and the small quantity of fleshy matter that exists on the outer extremity of the wings of birds; the former affording an obstacle, and the latter no temptation to the hyænas to devour them. Two of the bones here mentioned (see Plate II. fig. 19 to 23), in size and form, and the position of the points at the base of the quills, exactly resemble the ulna of a raven; a third (fig. 26, 27) approaches closely to the Spanish runt, which is one of the largest of the pigeon tribe; a fourth bone (fig. 24, 25) is the right ulna of a lark; a fifth (fig. 28, 29) the cora-

---

* The teeth and bones of water rats have been found by M. Cuvier to occur abundantly in many of the osseous breccias from the shores of the Mediterranean and Adriatic. He has also in his collection a large mass from Sardinia, composed exclusively of the bones and teeth of these animals, nearly as white as ivory, and slightly adhering together by delicate stalagmite; but by what process these bones were collected together, and whether in the antediluvian period, or more recently, it is not possible to decide, without careful examination of the spots in which they are respectively found, unless they happen to be in the same mass with bones or teeth belonging to extinct species.

coid process of the right scapula of a small species of duck resembling the Anas sponsor, or summer duck; and a sixth (Plate XIII. fig. 11, 12), has not been ascertained; it is nearly of the size of a thrush*.

With respect to the bear and tiger, the remains of which are extremely rare, and of which the teeth that have been found (see Plate VI. fig. 1, 5, 6, and 7), indicate a magnitude equal to the great Ursus spelæus of the caves of Germany, and of the largest Bengal tiger, it is more probable that the hyænas found their dead carcasses, and dragged them to the den, than that they were ever joint tenants of the same cavern. It is however obvious that they were all at the same time inhabitants of antediluvian Yorkshire.

In the case of such minute and burrowing animals as the mouse and weasel, and perhaps the rabbit and fox, it is possible that some of them may have crept into the cave by undiscovered crevices, and there died since the stoppage of its mouth; and in such case, their bones would have been found lying on the surface of the mud before it was disturbed by digging: as no observations were made in season as to this point, it must remain unsettled, till the opening of another cave may give opportunity for more accurate investigation. This uncertainty, however, applies not to any of the extinct species, or to the larger animals, whose habit it is not to burrow in the ground, nor even to those of the smaller ones, (e. g. the water rat,) fragments of whose bones and teeth are found imbedded in the antediluvian

* For my knowledge of these, and many other bones I have from Kirkdale, I am indebted to a careful examination and comparison of them made by Mr. Brooks, in his most valuable collection of osteological preparations. Mr. Clift also has kindly assisted me at the Royal College of Surgeons in furtherance of the same object.

stalagmite, and cemented by it both to the exterior and internal cavities of bones belonging to the hyænas and other extinct species, which, beyond all doubt, were lodged in the den before the period of the introduction of the mud. Should it turn out that since this period the cave has been accessible to foxes and weasels, it is possible that some of the birds also may have been introduced by them. The evidence of this, however, rests on a fact not yet carefully ascertained, viz. whether the bones in question were buried, like those of the extinct animals, beneath the mud, or lay on its surface; the state of one of the ravens' bones, containing stalagmite in its central cavity (see Plate XI. fig. 22, 23), seems to indicate high antiquity; and the quarryman, who was the first to enter the cave, assured me, that he has never seen a single bone of any kind on the surface, nor without digging into the substance of the mud, with the exception of a very few such cases as that of the specimen I have described in the possession of Mr. Smith.

As ruminating animals form the ordinary food of beasts of prey, it is not surprising that their remains should occur in such abundance in the cave (see Plate VIII. fig. 1 to 14); but it is not so obvious by what means the bones and teeth of the elephant, rhinoceros, and hippopotamus, were conveyed thither (see Plate VII. fig. 1 to 6, and 8 to 10). On the one hand, the cave is in general of dimensions so contracted (often not exceeding three feet in diameter), that it is impossible that living animals of these species could have found an entrance, or the entire carcasses of dead ones been floated into it; moreover, had the bones been washed in, they would probably have been mixed with pebbles and rounded equably by friction, which they are not: on the other hand, it is foreign to the habits of the hyæna to prey on the larger pachydermata, their young perhaps excepted. No other solution of the difficulty presents itself to me, than that the remains in question are those of individuals that died a natural death; for though an hyæna would neither have had strength to kill a living elephant or rhinoceros, or to drag home the entire carcass of a dead one, yet he could carry away, piecemeal, or acting conjointly with others, fragments of the most bulky animals that died in the course of nature, and thus introduce them to the inmost recesses of his den.

Should it be asked why, amidst the remains of so many hundred animals, not a single skeleton of any kind has been found entire, we see an obvious answer, in the power and known habit of hyænas to devour the bones of their prey; and the gnawed fragments on the one hand, and album græcum on the other, afford double evidence of their having largely gratified this natural propensity: the exception of the teeth and numerous small bones of the lower joints and extremities, that remain unbroken, as having been too hard and solid to afford inducement for mastication, is entirely consistent with this solution*.   And should it be further asked, why we do not find, at

---

* Since this paper was first published, I have had an opportunity of seeing a Cape hyæna at Oxford, in the travelling collection of Mr. Wombwell, the keeper of which confirmed in every particular the evidence given to Dr. Wollaston by the keeper at Exeter 'Change.   I was enabled also to observe the animal's mode of proceeding in the destruction of bones: the shin bone of an ox being presented to this hyæna, he began to bite off with his molar teeth large fragments from its upper extremity, and swallowed them whole as fast as they were broken off.   On his reaching the medullary cavity, the bone split into angular fragments, many of which he caught up greedily and swallowed

least, the entire skeleton of the one or more hyænas that died last and left no survivors to devour them; we find a sufficient reply to this question, in the circumstance of the probable destruction of the last individuals by the diluvian waters: on the rise of these, had there been any hyænas in the den, they would have rushed out, and fled for safety to the hills; and if absent, they could by no possibility have returned to it from the higher levels: that they were extirpated by this catastrophe is obvious, from the discovery of their bones in the diluvial gravel both of England and Germany.   The same circum-

entire: he went on cracking it till he had extracted all the marrow, licking out the lowest portion of it with his tongue: this done, he left untouched the lower condyle, which contains no marrow, and is very hard.   The state and form of this residuary fragment are precisely like those of similar bones at Kirkdale; the marks of teeth on it are very few, as the bone usually gave off a splinter before the large conical teeth had forced a hole through it; these few, however, entirely resemble the impressions we find on the bones at Kirkdale; the small splinters also in form and size, and manner of fracture, are not distinguishable from the fossil ones.   I preserve all the fragments and the gnawed portions of this bone for the sake of comparison by the side of those I have from the antediluvian den in Yorkshire: there is absolutely no difference between them, except in point of age.   The animal left untouched the solid bones of the tarsus and carpus, and such parts of the cylindrical bones, as we find untouched at Kirkdale, and devoured only the parts analogous to those which are there deficient.   The keeper pursuing this experiment to its final result, presented me the next morning with a large quantity of album græcum, disposed in balls, that agree entirely in size, shape, and substance with those that were found in the den at Kirkdale.   I gave the animal successively three shin bones of a sheep; he snapped them asunder in a moment, dividing each in two parts only, which he swallowed entire, without the smallest mastication.   On the keeper putting a spar of wood, two inches in diameter, into his den, he cracked it in pieces as if it had been touchwood, and in a minute the whole was reduced to a mass of splinters.   The power of his jaws far exceeded any animal force of the kind I ever saw exerted, and reminded me of nothing so much as of a miner's crushing mill, or the scissars with which they cut off bars of iron and copper in the metal founderies.

stance will also explain the reason why there are no heaps of bones found on the outside of the Kirkdale cave, as described by Busbequius on the outside of the hyænas' dens in Anatolia; for every thing that lay without, on the antediluvian surface, must have been swept far away, and scattered by the violence of the diluvian waters; and there is no reason for believing that hyænas, or any other animals whatever, have occupied the den subsequently to that catastrophe*.

Although the evidence to prove the cave to have been inhabited as a den by successive generations of hyænas appears thus direct, it may be as well to consider what other hypotheses can be suggested, to explain the collection of bones assembled in it.

1st. It may be said, that the various animals had entered the cave spontaneously to die, or had fled into it as a refuge from some general convulsion: but the diameter of the cave, as has been mentioned before, compared with the bulk of the elephant and rhinoceros, renders this solution impossible as to the larger animals; and with respect to the smaller, we can imagine no circumstances that would collect together, spontaneously, animals of such dissimilar habits as hyænas, tigers, bears, wolves, foxes, horses, oxen, deer, rabbits, water-rats, mice, weasels, and birds.

2d. It may be suggested, that they were drifted in by the waters

---

* It has been suggested further, that there is no proof that this individual cave was actually occupied at the precise point of time at which the waters began to rise, although it certainly had been so during several generations not long preceding. It may have been abandoned a short time prior to it, and at that moment have been untenanted; for modern hunters do not always find their game exactly on the same spot, nor is there any thing to prevent hyænas as well as other wild animals from occasionally changing their quarters. *Quarterly Review,* Oct. 1822, p. 468.

of a flood: if so, either the carcasses floated in entire; or the bones alone were drifted in after separation from the flesh: in the first of these cases, the larger carcasses, as we have already stated, could not have entered at all; and of the smaller ones, the cave could not have contained a sufficient number to supply one-twentieth part of the teeth and bones; moreover, the bones would not have been broken to pieces, nor in different stages of decay. And had they been washed in by a succession of floods, we should have had a succession of beds of sediment and stalactite, and the cave would have been filled up by the second or third repetition of such an operation as that which introduced the single stratum of mud, which alone occurs in it. On the other hypothesis, that they were drifted in after separation from the flesh, they would have been mixed with gravel, and at least slightly rolled on their passage; and it would still remain to be shown by what means they were split and broken to pieces, and the disproportion created which exists between the numbers of the teeth and bones. They could not have fallen in through the fissures, for these are closed upwards in the substance of the rock, and do not reach to the surface.

The 3rd, and only remaining hypothesis that occurs to me is, that they were dragged in for food by the hyænas, who caught their prey in the immediate vicinity of their den; and as they could not have dragged it home from any very great distances, it follows, that the animals they fed on all lived and died not far from the spot where their remains are found.

The accumulation of these bones, then, appears to have been a long process, going on during a succession of years, whilst all the animals in question were natives of this country. The general dispersion of bones of the same animals through the diluvian gravel of high latitudes, over great part of the northern hemisphere, shows that the period in which they inhabited these regions was that immediately preceding the formation of this gravel, and that they perished by the same waters which produced it. M. Cuvier has moreover ascertained, that the fossil elephant, rhinoceros, hippopotamus, and hyæna, belong to species now unknown; and as there is no evidence that they have at any time, subsequent to the formation of the diluvium, existed in these regions, we may conclude that the period, at which the bones of these extinct species were introduced into the cave at Kirkdale, was antediluvian. Had these species ever reestablished themselves in the northern portions of the world since the deluge, it is probable their remains would have been found, like those of the ox, horse, deer, hog, &c. preserved in the postdiluvian accumulations of gravel, sand, silt, mud, and peat, which are referable to causes still in operation, and which, by careful examination of their relations to the adjacent country, can be readily distinguished from those which are of diluvian origin.

The teeth and fragments of bones above described seem to have lain a long time scattered irregularly over the bottom of the den, and to have been continually accumulating until the introduction of the sediment in which they are now imbedded, and to the protection of which they owe that high state of preservation they possess. Those that lay long uncovered at the bottom of the den have undergone a decay proportionate to the time of their exposure; others, that have

G

lain only a short time before the introduction of the diluvian mud, have been preserved by it almost from even incipient decomposition.

Thus the phenomena of this cave seem referable to a period immediately antecedent to the last inundation of the earth, and in which the world was inhabited by land animals, almost all bearing a generic and many a specific resemblance to those which now exist; but so completely has the violence of that tremendous convulsion destroyed and remodelled the form of the antediluvian surface, that it is only in caverns that have been protected from its ravages that we may hope to find undisturbed evidence of events in the period immediately preceding it. The bones already described, and the stalagmite formed before the introduction of the diluvial mud, are what I consider to be the products of the period in question. It was indeed probable, before the discovery of this cave, from the abundance in which the remains of similar species occur in superficial gravel beds, which cannot be referred to any other than a diluvial origin, that such animals were the antediluvian inhabitants not only of this country, but generally of all those northern latitudes in which their remains are found (but the proof was imperfect, as it was possible they might have been drifted or floated hither by the waters from the warmer regions of the earth); but the facts developed in this charnel-house of the antediluvian forests of Yorkshire demonstrate that there was a long succession of years in which the elephant, rhinoceros, and hippopotamus had been the prey of the hyænas, which, like themselves, inhabited England in the period immediately preceding the formation of the diluvial gravel; and if they inhabited this country, it follows as a corollary, that they also inhabited all those other regions of the northern hemisphere in which similar bones have been found under precisely the same circumstances, not mineralised, but simply in the state of grave bones imbedded in loam, or clay, or gravel, over great part of northern Europe, as well as North America and Siberia. The catastrophe producing this gravel appears to have been the last event that has operated generally to modify the surface of the earth, and the few local and partial changes that have succeeded it, such as the formation of deltas, terraces, tufa, torrent-gravel and peat-bogs, all conspire to show, that the period of their commencement was subsequent to that at which the dilivium was formed *.

---

* It was stated in describing the locality of the cave at Kirkdale, and on comparing it with the fact of its containing the remains of large and small aquatic animals, that there was probably a lake in this part of the country at the period when they inhabited it; and this hypothesis is rendered probable by the form and disposition of the hills that still encircle the Vale of Pickering. (See Map, Plate I.)

Inclosed on the south, the west, north-west, and north, by the lofty ranges of the Wolds, the Howardian hills, the Hambleton hills, and Eastern Moorlands, the waters of this vale must either run eastward to Filey Bay, or inland towards York; and such is the superior elevation of the strata along the coast, that the sources of the Derwent, rising almost close to the sea, near Scarborough and Filey, are forced to run west and southward fifty miles inland away from the sea, till falling into the Ouse, they finally reach it by turning again eastward through the Humber. The only outlet by which this drainage is accomplished is the gorge at New Malton; and though it is not possible to ascertain what was the precise extent of this antediluvian lake, or how much of the low districts, now constituting the Vale of Pickering, may have been excavated by the same diluvian waters that produced the gorge, it is obvious, that without the existence of this gorge, much of the district within it would be laid under water; and it is not till within these few years that a large tract of this land has been recovered from a state of swamp and marsh by an artificial canal, called the Muston Drainage, which runs inland from the sea westward along the valley of the Derwent, from Muston, near Filey Bay, to the gorge of New Malton; it is equally obvious, that this gorge is referable to the agency of diluvial denudation, the ravages of which have not, perhaps, left a single portion of

It is in the highest degree curious to observe, that four of the genera of animals whose bones are thus widely diffused over the temperate, and even polar regions of the northern hemisphere, should at present exist only in tropical climates, and chiefly south of the equator; and that the only country in which the elephant, rhinoceros, hippopotamus, and hyæn aare now associated is Southern Africa. In the immediate neighbourhood of the Cape they all live and die together, as they formerly did in Britain; whilst the hippopotamus is now confined exclusively to Africa, and the elephant, rhinoceros, and hyæna are also diffused widely over the continent of Asia.

To the question which here so naturally presents itself, as to what might have been the climate of the northern hemisphere when peopled with genera of animals which are now confined to the warmer regions of the earth, it is not essential to the point before me to find a solution; my object is to establish the fact, that the animals lived and died in the regions where their remains are now found, and were not drifted thither by the diluvian waters from other latitudes. The state of the climate in which these extinct species may have

the antediluvian surface of the whole earth, which is not excavated and re-modelled, so as to have lost all traces of the exact features it bore antecedently to the operations of the deluge.

It is probable, that inland lakes were much more numerous than they are at present, before the excavation of the many gorges by which our modern rivers make their escape; and this is consistent with the frequent occurrence of the remains of the hippopotamus in the diluvian gravel of England, and of various parts of Europe, particularly in the Val d'Arno. It is not unlikely that, in this antediluvian period, England was connected with the Continent, and that the excavation of the shallow channel of the Straits of Dover, and of a considerable portion of that part of the German ocean which lies between the east coast of England and the mouths of the Elbe and Rhine, may have been the effect of diluvial denudation. The average depth of all this tract of water is said to be less than thirty fathoms.

lived antecedently to the great inundation by which they were extirpated is a distinct matter of inquiry, on which the highest authorities are by no means agreed. It is the opinion of Cuvier, on the one hand, that as some of the fossil animals differ from existing species of the genera to which they belong, it is probable they had a constitution adapted to endure the rigours of a northern winter; and this opinion derives support from the Siberian elephant's carcase, discovered with all its flesh entire, in the ice of Tungusia, and its skin partially covered by long hair and wool; and from the hairy rhinoceros found in 1771 in the same country, in the frozen gravel of Vilhoui, having its flesh and skin still perfect, and of which the head and feet are now preserved at Petersburg, together with the skeleton of the elephant above alluded to, and a large quantity of its wool; to which Cuvier adds the further fact, that there are genera of existing animals, e. g. the fox tribe, which have species adapted to the extremes both of polar and tropical climates.

On the other hand, it is contended that the abundant occurrence of fossil crocodiles and tortoises, and of vegetables and shells (e. g. the nautilus), nearly allied in structure and character to those which are now peculiar to hot climates, in the secondary strata, as well as in the diluvium of high north latitudes, renders it more probable that the climate was warm in which these plants and animals lived and died, than that a change of constitution and habit should have taken place in so many animal and vegetable genera, the existing members of which are rarely found except in the warmer regions of the present earth. To this argument, I would add a still greater objection arising from the difficulty of maintaining such animals as those we are considering amid the rigours of a polar winter; and this difficulty cannot be solved by supposing them to have migrated periodically, like the musk ox and rein deer of Melville Island; for in the case of crocodiles and tortoises extensive emigration is almost impossible, and not less so to such an unwieldy animal as the hippopotamus when out of water. It is equally difficult to imagine that they could have passed their winters in lakes or rivers frozen up with ice; and though the elephant and rhinoceros, if clothed in wool, may have fed themselves on branches of trees and brushwood during the extreme severities of winter, still I see not how even these were to be obtained in the frozen regions of Siberia, which at present produce little more than moss and lichens, which during great part of the year are buried under impenetrable ice and snow; yet it is in these regions of extreme cold, on the utmost verge of the now habitable world, that the bones of elephants are found occasionally crowded in heaps along the shores of the icy sea from Archangel to Behring's Straits, forming whole islands composed of bones and mud at the mouth of the Lena, and encased in icebergs, from which they are melted out by the solar heat of their short summer, along the coast of Tungusia, in sufficient numbers to form an important article of commerce *.

* " Lieutenant Kotzebue has discovered, in the western part of the gulf to the north of Behring's Straits, a mountain covered with verdure (moss and grass) composed interiorly of solid ice. On arriving at a place where the shore rises almost perpendicularly from the sea to the height of 100 feet, and continues afterwards to extend with a gradual inclination, they observed masses of the purest ice 100 feet high, preserved under the above vegetable carpet. The portion exposed to the sun was melting and sending much water into the sea. An undoubted proof of this ice being primitive (i. e. not formed by any causes now in action), was afforded by the great number of bones and teeth of mammoths which make their appearance when it is melted. The soil of these

Between these two conflicting opinions we are compelled to make our choice: there seems to be no third or intermediate state with which both may be compatible. It is not, however, to my present purpose to discuss the difficulties that will occur on both sides, till the further progress of geological science shall have afforded us more ample information as to the structure of our globe, and have supplied those data, without which all opinions that can be advanced on the subject must be premature, and amount to no more than plausible conjecture. At present I am concerned only to establish two important facts, 1st, that there has been a recent and general inundation of the globe; and, 2d, that the animals whose remains are found interred in the wreck of that inundation were natives of high north latitudes, and not drifted to their present place from equatorial regions by the waters that caused their destruction. One thing, however, is nearly certain, viz. that if any change of climate has taken place, it took place suddenly; for how otherwise could the elephant's carcase, found entire in ice at the mouth of the Lena, have been preserved from putrefaction till it was frozen up with the waters of the then existing ocean? Nor is it less probable that this supposed change was contemporaneous with, and produced by, the same cause which brought on the inundation. What this cause was, whether a change in the inclination in the earth's axis, or the near approach of a comet, or any other cause or combination of causes purely astrono-

mountains, which, to a certain height, are covered with an abundant herbage, is only half a foot thick; it is composed of a mixture of clay, earth, sand, and mould; the ice melts gradually beneath it, the carpet falls downwards and continues to thrive; the latitude is 66° 15' 36" N."—*Gilbert's* Annalen, 1821, quoted in the Journal of Science and the Arts, No. 27, page 236.

mical, is a question the discussion of which is foreign to the object of the present memoir.

Having thus far described the principal facts I observed in the interior of the den at Kirkdale, and pointed out the most important conclusions that seem to arise from them, I proceed to consider the chronological inferences that may be derived from the state of the bones, and of the mud and stalagmite that accompany them, and to extract the following detail of events that have been going on successively within this curious cave.

1st. There appears to have been a period (and if we may form an estimate from the small quantity of stalagmite now found on the actual floor of the cave, a very short one,) during which this aperture in the rock existed in its present state, but was not tenanted by the hyænas. The removal of the mud, which now entirely covers the floor, would be necessary to ascertain the exact quantity of stalagmite referable to this period, but it cannot be very great, and can only be expected to exist where there is much stalactite also upon the roof and sides.

The 2d period was that during which the cave was inhabited by the hyænas, and the stalactite and stalagmite were still forming. The constant passage of the hyænas in so low a cave would much interrupt this deposition; as they would strike off the former from the roof and sides by their constant ingress and egress; and accordingly in some specimens of the breccia, we find mixed with the bones fragments of stalactite, that seem to have been thus knocked off from the roof and sides of the cave, whilst it was inhabited by hyænas before the introduction of the mud. I have one example of a hollow stalactitic tube that lay in an horizontal position in the midst of, and parallel to some splinters of large bones, and the unbroken ulna of a rat: all these are united by stalagmite; and it is impossible that this stalactitic pipé could have been formed in any other than a vertical position, hanging from the roof or sides. In other specimens of the breccia, there are split fragments of the teeth of deer and hyæna; and in almost every portion I have seen, either of this breccia or of the antediluvian stalagmite, there are teeth of the water rat. Mr. Gibson possesses a mass exceeding a foot in diameter, composed of fragments of many large bones, mixed with some teeth of rhinoceros and several of the larger animals, and also of rats, all adhering firmly together in a matrix of stalagmite. During the formation of the stalagmitic matter, no mud appears to have been introduced; and had there been any in the cave at the time whilst the osseous breccia was forming, it would either have excluded all access of the stalagmite to the bones, or have been mixed and entangled with it, forming a spongy mass, as it does at the root of the stalagmites that lie on its surface. The universal cover of mud prevented me from ascertaining whether the bottom of the cave is any where polished (like the tiger's den before alluded to,) in those parts which must have been the constant gangway of the hyænas.

The 3d period is that at which the mud was introduced and the animals extirpated, viz. the period of the deluge. I have already stated that the animal remains are found principally in the lower regions of this sediment of mud, which appears to have been intro- duced in a fluid state, so as to envelope the bony fragments then

H

lying on the bottom of the cave : and the power of water to introduce such sediments is shown by the state of Wokey Hole, and similar caverns in the Mendip Hills, and Derbyshire, which are subject to be filled with water occasionally by heavy land floods. The effect of these floods being to leave on the floor a sediment of mud similar to that which covers the bones and osseous breccia in the cave of Kirkdale.   I have also mentioned that there is no alternation of this mud with beds of bone or of stalagmite, such as would have occurred had it been produced by land floods often repeated ; once, and once only, it appears to have been introduced ; and we may consider its vehicle to have been the turbid waters of the same inundation that produced universally the diluvial gravel and loam on the surface without : these would enter and fill the cave, and there become quiescent, would deposit the mud suspended in them (as we see daily silt and warp deposited in quiet spots by waters of muddy rivers) along the whole bottom of the den, where it has remained undisturbed ever since. We cannot refer this mud to a land flood, or a succession of land floods, partly for the reasons before stated, and partly from the general dryness of the cave ; had it been liable to be filled with muddy water, it would have been so at the time I visited it in December, 1821, at the end of one of the most rainy seasons ever remembered ; but even then there were not the slightest symptoms of any such occurrence, and a few scanty droppings from the roof were the only traces of water entering the area of the cave.

The 4th period is that during which the stalagmite was deposited which invests the upper surface of the mud. The quantity of this stalagmite appears to be much greater than that formed in the two periods, during and before which, the cave was tenanted by hyænas. In the whole of this 4th period no creature appears to have entered the cave, with the exception possibly of mice, rats *, weasels, rabbits, and foxes, until it was opened last summer; and no other process of any kind appears to have been going on in it except the formation of stalactitic and stalagmitic infiltrations: the stratum of diluvial sediment marks the point of time at which the latter state of things began and the former ceased. As there is no mud at all on the top or sides of the cave, we have no mark to distinguish the relative quantities of stalactite formed on these parts during the periods we have been speaking of: should it however contain in any part a fragment of bone or tooth of any of the extinct animals, it will be probable that this part was antediluvial. A farther argument may be drawn from the limited quantity of postdiluvian stalactite, as well as from the undecayed condition of the bones, to show that the time elapsed since the introduction of the diluvial mud has not been one of excessive length, nor at all exceeding that which M. Cuvier, after comparing the traditions of a deluge that prevail among all nations with natural phenomena, infers to have elapsed since that great and universal inundation which has overwhelmed the earth, at a period which, he says, he is of opinion with De Luc and Dolomieu, cannot have exceeded five or six thousand years ago.

---

* Mr. Salmond has a portion of the upper stalagmite, with the entire skeleton of a rat, embedded between two of the upper laminæ of the stalagmitic crust. This animal must have entered the cave, and died there, not long ago.

CAVES AT KIRBY MOORSIDE.

I mentioned in my former paper, that a second cave had been discovered in the vicinity of Kirkdale, which was reported also to contain bones, and that it had been closed by Mr. Duncombe till I should come down to examine it, which I did in July last, accompanied by Sir H. Davy and Mr. Warburton. Our labour was lost as far as related to the discovery of more bones, or a second den of hyænas; but it was repaid by the confirmation which this cave afforded in all its other circumstances of my speculations on that at Kirkdale, and by the discovery of another cavity in Duncombe park containing animal remains, which throw much light on the mode in which the caves and fissures that were not inhabited as dens became filled with bones. I had also the satisfaction of demonstrating on the spot to Sir H. Davy and Mr. Warburton the actual state of many of the phenomena described in my account of Kirkdale. The cave at Kirby Moorside was intersected in working the face of a quarry of the same limestone as that at Kirkdale, at the north end of the town, and on the right side of a narrow gorge or valley called the Manor Vale, which descends from the north towards the Vale of Pickering, nearly parallel to the valley of Kirkdale, being about sixty feet broad, and bounded by slopes forty feet high, and forming one of the many smaller vallies of denudation excavated on this limestone by the diluvial waters as they subsided from the moorlands to the Vale of Pickering. A considerable portion of the right bank of this valley has been laid bare by the workings of the quarry, and on the face of it there are traces of a fissure connected with several small cavernous holes. The aperture discovered last spring is in the centre of this quarry, and near its floor; on removing the wall with which Mr. Duncombe had caused it to be closed, it was found to pass obliquely into the body of the hill, and to be intersected at a few feet from its entrance by a large fissure; this point of intersection forms, as at Kirkdale, the widest and most lofty part of the cavern, within which it diminishes into smaller vaults, which soon become impassable: the outer part of the cave when first opened was about four feet high and six broad, and its entire floor covered with an uniform mass of loamy clay, precisely similar to that on the floor of the den at Kirkdale. On digging into this loam it was found to be six feet deep for a considerable distance inwards, and to contain no bones. At its bottom there was no stalagmitic undercrust dividing it from the limestone floor, nor any repetition or alternation of a second or third bed of stalagmite in any part of its substance; its surface alone was in many parts glazed over with an extensive sheet of it oozing outwards from the side walls, and sometimes entirely crossing and forming a bridge over the loam. Above this crust some parts of the roof and sides were loaded with stalactite in its usual fantastic forms; but there were no bones of modern animals, nor traces of loam, or even of dust, upon the surface of the superficial crust of stalagmite. In all its circumstances, as far as they went, it agreed with and confirmed the history and chronology I have given of the cave at Kirkdale, excepting the two accidents of its not having been inhabited as a den, or received any stalagmite on its floor, before the introduction of the diluvial loam. The absence of bones in this cave (the mud being present) adds to the probability that it was by the instrumentality of the hyænas, and not of the diluvial waters, that the animal remains were collected in such quantities in the adjacent den at Kirkdale.

At about a mile east of Kirby Moorside, at a spot called the Back of the Parks, there are other quarries on both sides of a comb that descends rapidly into the valley of the Dove, in the face of which there occur several small caverns and vertical fissures: these fissures vary from one to six feet in breadth, and rise from the bottom of the quarry to the surface of the land, and are entirely filled with diluvial loam, of the same kind as that in the caves both here and at Kirkdale, and the Manor Vale.   It was in the upper part of one of the fissures that several human skeletons were found and taken out in the year 1786, but the spot on which they occurred has been destroyed in continuing the workings of the quarry: they were probably bodies that had been interred here after a battle.

OPEN FISSURE IN DUNCOMBE PARK.

The newly discovered fissure in Duncombe Park differs from those we have been last describing in the circumstance of its being of post-diluvian origin; it contains no diluvial sediment and no pebbles, and has within it the remains of animals of existing species only, and these in a much more recent and more perfect state of preservation than the bones at Kirkdale.   It is a great irregular crack or chasm, in the solid limestone rock, which forms a steep and lofty cliff on the right side of the valley of the Rye, being in that most beautiful valley of denudation which descends from Rivaulx Abbey through Duncombe

Park to the town of Helmsley, and on the left bank of which are the magnificent terraces of Rivaulx Abbey, and of the gardens at Duncombe Park. The crack has probably been formed by a subsidence of part of the cliff towards the valley, and terminates upwards near its edge, in a small aperture, about twenty feet long and three or four feet broad, which is almost concealed and overgrown with bushes, and which being nearly at right angles to the edge of the cliff, lies like a pitfall across the path of animals that pass that way. It descends obliquely downwards, and presents several ledges or landing places and irregular lateral chambers, the floors of which are strewed over with loose angular fragments of limestone, fallen from the sides and roof, and with dislocated skeletons of animals that have from time to time fallen in from above and perished. One of Mr. Duncombe's game-keepers had been for many years aware of the existence of bones in this chasm, but had never mentioned it till my second visit to Duncombe Park, when we examined it, descending by means of a rope, and found it to contain the skeletons of dogs, sheep, deer, goats, and hogs, lodged at various depths on the landing places I have just mentioned: the bones lay loose and naked on the actual spots on which the animals had died, and to which they had probably fallen when passing carelessly along the surface of the Park above; they were neither broken, nor buried in loam, nor incrusted with stalagmite, as at Kirkdale, but simply stripped of their flesh; they are not adherent to the tongue when fractured, but retain much more animal matter, and are in all respects more fresh and recent, than those which occur at Kirkdale entombed beneath the loam.

In a geological point of view, the occurrence of these bones, under the circumstances above described, is important, as illustrating the manner in which the bones of antediluvian animals may have been accumulated by falling into similar fissures, which are now filled up with diluvial mud and pebbles; for if fissures existed (as they undoubtedly did) on the antediluvian face of the earth in much greater abundance than since that grand aqueous revolution, which has entirely filled up so many of them with its detritus, there is no reason why the then existing animals should not have fallen into them and perished, as modern animals do in the comparatively few cavities that remain still open in our limestone districts: and when we consider that it is the habit of graminivorous animals to be constantly traversing the surface of the ground in every direction in pursuit of food, it is obvious that they are subject in a greater degree than those which are carnivorous to the perpetual danger of falling into any fissure or imperfectly closed chasm that may lie in their way; and in this circumstance we see an explanation of the comparatively rare occurrence of the remains of beasts of prey in the osseous breccia of the antediluvian fissures, although they also occasionally perished in them, as the dogs do at this day in the open fissure at Duncombe Park.

Many of the arguments arising from the detail of facts we have been describing in Yorkshire are applicable to the illustration of analogous phenomena, where the evidence of their history is less complete. In our own country there are seven other instances of bones similarly deposited in caverns, the origin of some of which, though not before satisfactorily made out, becomes evident as a corollary from the proofs afforded by the cave at Kirkdale; these are in the counties of Somerset, Derby, Devon, and Glamorganshire.

1.—CAVE OF HUTTON, IN THE MENDIP HILLS.

The first I shall mention is that of teeth and bones of elephants and other antediluvian animals discovered in the Mendip Hills in cavities of mountain limestone, which were lined, and nearly filled with ochreous clay.   These are preserved in the collection of the Rev. Mr. Catcott, in the City Library at Bristol.   The following account of them is extracted by my friend the Rev. W. D. Conybeare, from Mr. Catcott's MS. notes; he has added also a few explanatory observations.

" The ochre pits were worked about the middle of the last century, near the summit of the Mendip Hills, on the S. of the village of Hutton, near Banwell, at an elevation of from three hundred to four hundred feet above the level of the sea: they are now abandoned*.

" The ochre was pursued through fissures in the mountain limestone, occasionally expanding into larger cavernous chambers, their range being in a steep descent, and almost perpendicular.   Thus, in opening the pits, the workmen, after removing eighteen inches of vegetable mould, and four feet of rubbly ochre, came to a fissure in the limestone rock, about eighteen inches broad, and four feet long. This was filled with good ochre, but as yet no bones were discovered:

* I shall presently mention an analogous case of the occurrence of ochre in a similar series of caverns in Derbyshire, near Wirksworth, and in some caves and fissures, filled with a similar accumulation of diluvial matter, on the continent, at Theux, near Spa. In the latter case it is accompanied with a large admixture of pebbles, but no bones.

I

it continued to the depth of eight yards, and then opened into a cavern about twenty feet square, and four high; the floor of this cave consisted of good ochre, strewed on the surface of which were multitudes of white bones, which were also found dispersed through the interior of the ochreous mass. In the centre of this chamber, a large stalactite depended from the roof; and beneath, a similar mass rose from the floor, almost touching it: in one of the side walls was an opening about three feet square, which conducted through a passage eighteen yards in length, to a second cavern, ten yards in length, and five in breadth; both the passage and cavern being filled with ochre and bones. Another passage, about six feet square, branched off laterally from this chamber about four yards below its entrance; this continued nearly on the same level for eighteen yards; it was filled with rubbly ochre, fragments of limestone rounded by attrition, and lead ore confusedly mixed together; many large bones occurring in the mass; among which four magnificent teeth of an elephant (the whole number belonging to a single skull) were found. Another shaft was sunk from the surface perpendicularly into this branch, and appears to have followed the course of a fissure, since it is said that all the way nothing appeared but rubble, large stones, ochre, and bones: in the second chamber, immediately beyond the entrance of the branch just described, there appeared a large deep opening, tending perpendicularly downwards, filled with the same congeries of rubble, ochre, bones, &c.; this was cleared to the depth of five yards; this point, being the deepest part of the workings, was estimated at about thirty-six yards beneath the surface of the hill; a few yards to the west of this another similar hole occurred, in which was found a large head, which we shall have occasion presently to notice."

The bones from this cavern, preserved in Mr. Catcott's cabinet in the Bristol library, are the teeth and fragments of some bones of the elephant; and similar remains of horses, oxen, and two species of stag, besides the skeleton, nearly complete, of a fox, and the metacarpal bone of a very large species of bear, nearly five inches in length. There are also molar teeth of the hog, and a large tusk of the upper jaw; (see Plate XI. fig. 30, 31, 32, 33.) This tusk probably belonged to the head mentioned in his MS. as having been found in the pit above described, and of which the following particulars are specified:—" The head was stated by the workmen to have been about three or four feet long, fourteen inches broad at the top, or head part, and three inches at the snout. It had all the teeth perfect, and four tusks, the larger tusks about four inches long out of the head, and the lesser about three inches*." The tusk now preserved is about three inches long, its enamel is fine, it is longitudinally striated, and on one side of the apex truncated and worn flat by use.

Some farther details of the bones found in the cave at Hutton are given as a note in Mr. Catcott's Treatise on the Deluge (page 361, first edition), in which he specifies six molar teeth of the elephant, one of them lying in the jaw, part of a tusk, part of a head, four thigh bones, three ribs, with a multitude of lesser bones, belonging probably to the same animal. " Besides these," he adds, " we picked up part

---

* The head here described is evidently that of a hog; the account of its length being exaggerated by the workmen, from whose report alone Mr. Catcott gives the measures of it. The head itself was lost or destroyed before he had seen it.

of a large deer's horn very flat, and the slough of a horn (or the spongy porous substance that occupies the inside of the horns of oxen), of an extraordinary size, together with a great variety of teeth and small bones belonging to different species of land animals. The bones and teeth were extremely well preserved, all retaining their native whiteness, and, as they projected from the sides and top of the cavity, exhibited an appearance not unlike the inside of a charnel-house."

It appears most probable, from the description given of these bones and horns, that they were not dragged in by beasts of prey, but either drifted in by the diluvian waters, or derived from animals that had fallen in before the introduction of the ochreous loam: the loam itself and pebbles are clearly of diluvial origin.

On the summit of Sandford Hill, on the east of Hutton, bones of the elephant were also, according to Mr. Catcott's MSS., discovered four fathoms deep among loose rubble.

2.—CAVE ON DERDHAM DOWN, NEAR CLIFTON.

A second case of fossil fragments of bone has been discovered by Mr. Miller, of Bristol, in a cavity of mountain limestone, near Clifton, by the turnpike-gate on Derdham Down: these are not rolled, but have evidently been fractured by violence: they are partially incrusted with stalactitic matter, and the broken surfaces have also an external coating of thin ochreous stalactite, showing the fracture to have been ancient. One specimen, the property of Mr. Miller, displays the curious circumstance of a fossil joint of the horse; it is the tarsus joint, in which the astragalus retains its natural position between the tibia and os calcis; these are held together by a stalagmitic cement, and were probably left in this position by some beast of prey that had gnawed off the deficient portions of the tibia and os calcis.

3.—CAVE AT BALLEYE, NEAR WIRKSWORTH.

A third case is that of some bones and molar teeth of the elephant, found in another cavity of mountain limestone at Balleye, near Wirksworth, in Derbyshire, in the year 1663; one of these teeth is now in the collection of Mr. White Watson, of Bakewell. There is, I believe, no detailed account of the circumstances under which these remains were found, farther than that the cavity was intersected in working a lead mine.

4.—DREAM CAVE, NEAR WIRKSWORTH.

A fourth example has just occurred in the same neighbourhood, in a lead mine called the Dream, in the hamlet of Callow, about one mile W. of Wirksworth, towards Hopton, on the property of Philip Gell, Esq., whose attention has been judiciously directed to the subject, and by whose exertions it is probable that nearly the entire skeleton of a rhinoceros will be extracted and preserved, together with some remains of the ox and deer. On being informed of this discovery, through the kindness of my friend the Rev. D. Stacy, I set off immediately for Derbyshire, for the purpose of examining all its circumstances, and found them to be nearly as follows.   In the month of December last, 1822, some miners engaged in pursuing a lead vein had sunk a shaft about sixty feet through solid mountain lime-stone (see Plate XX. A. B. H.), when they suddenly penetrated a large cavern E, filled entirely to the roof with a confused mass of argilla-ceous earth and fragments of stone, through which they attempted to continue their shaft perpendicularly downwards to the vein below; in this operation they were interrupted by the earth and fragments beginning to move and fall in upon them continually from the sides, until the roof of a large cavern became apparent, in consequence of the subsidence and removal of the matter with which it had been filled.   It was nearly in the centre of this subsiding mass, and at the height of many feet above the actual floor of the cave, that the work-men discovered the bones F G, which I am about to describe, and of which those belonging to the rhinoceros G lay very near each other, and probably formed an entire skeleton before they were disturbed by the agitation and sinking of the materials E, in which they were imbedded.   The following parts of a rhinoceros, all apparently from the same individual, have already been collected; viz. the nose, part of the upper jaw, one large superior molar tooth, the entire back of the head, and one half of the under jaw, containing three perfect molar teeth; the atlas quite entire and fitting the head, two cervical vertebræ, two dorsal and one lumbar ditto, the sacrum and several parts of the pelvis, the humerus and ulna, fitting together, one femur, with its head fitting the acetabulum of the pelvis, the left patella, and several fragments of ribs.   All these are in a state of high preservation, equal to that of the bones from Kirkdale, and from a full grown animal, and coming as they do from such various parts of the body, and being found so close together, they leave no doubt but that they are portions of a skeleton which lay entire in the middle of the cave before the materials that had filled it began to subside, and of which it is probable that the remaining parts will on further search be also discovered. There were no supernumerary bones, to indicate the presence of a second rhinoceros, but simply the metatarsus of a very large ox, the radius and some other bones of a stag, and some portions of deer's horns, of the size but not the shape of those of a large fallow deer, and some of them having the upper extremities palmated. (See Plate XXII. fig. 3, 4.) None of these bones have marks of partial decay on one surface whilst the other remains entire, as happens at Kirkdale and Plymouth; and from this circumstance we may infer that they were derived from animals that perished by the same waters that introduced them to the cave*. For some time after the cave was penetrated there was no apparent communication between its interior and the upper surface; but as the loose materials that at first filled it subsided into and were taken out by the shaft, a gradual sinking appeared in the surface of the field above at ɪ, and a further mass of the same kind, viz. argillaceous earth and fragments of limestone, mixed with a few rolled pebbles of quartz, continued to fall downwards into it (like the contents of a lime-kiln, sinking towards the lower aperture by which the lime is extracted), until a large open chasm ᴅ, more than six feet broad, and fifty feet deep, was left en-

* Mr. Gell has also in his possession the horn of a very large urus, that was found at a considerable depth in digging away the diluvium near the west mouth of the tunnel of the Cromford Canal, at Butterley, about thirty years ago.

tirely void, and seen to form a direct communication from the side of the cave to the surface of the field above. Till undermined in this manner, the fissure D had been entirely filled, and the surface afforded not the slightest indication of its existence; at present it is restored to the same state of an open chasm in which it probably was before the access of the diluvian waters, that appear to have swept into it the mud and rocky fragments which filled both it and the cave below; and on examining its sides, I found the projecting parts of them rubbed and scratched by the descent of these heavy bodies as they dropped in from above. From the situation of the rhinoceros' bones in the middle of this drifted mass, and in the centre of the cave, added to the juxta-position of so many of the component parts of one entire skeleton, which are neither rolled, or gnawed, or broken, except by the move-ment they have recently undergone, and the pickaxes of the miners, it seems probable that they are the remains of a carcase that was drifted in entire at the same time with the diluvial detritus, in the midst of which they were found imbedded: had they been washed in singly, they would have been slightly rolled and scattered irregularly, and we should have found parts of more than a single individual; and had they been derived from an animal that fell into the fissure, and perished before the introduction of the diluvium, they would not have been suspended, as they were, all together nearly in the middle of it, but would have lain either on the actual floor of the cave be-neath the loam and pebbles, or have been scattered and drifted irre-gularly to different and distant parts of its lowest recesses. I could discover no stalagmite, and but few traces of stalactite in any part of this cavern, or of the fissure immediately connected with it.

In the same field with the Dream Mine, and on the upper edge of a steep declivity, is a small crag that overhangs the subjacent valley, and has in its face an aperture called the Fox Holes, by which we enter an extensive suite of connected chambers and vaultings of irregular size and shape, perforating the rock in various directions, and at various elevations. In all of these the floor is covered to the depth of many feet, whilst some of the smaller ones are entirely filled with a mass of clay and ochreous loam, which in many parts is sufficiently pure to have been extracted for sale as a coarse pigment, and to have caused much of the diluvial sediment within these chambers to have been dug over in search of it, as was done in the cavernous fissure of Hutton in the Mendips, just described as having contained the bones of an elephant and other animals imbedded in ochre. In the cave of Fox Holes, now before us, no bones have been discovered, nor are there any traces of pebbles or angular fragments of stone except near the mouth. The quantity of stalactite and stalagmite also is small: and this little occurs chiefly near the entrance, where the roof is clustered with tufts of beautiful lac lunæ. The position of the cave on the edge of a high cliff, and far above the possible influence of any floods from the nearest brooks or rivulets, obliges us to refer the enormous deposit it contains of ochreous mud to no other than diluvial origin; and Mr. Gell informs me, that in all the caves and in the greater number of the fissures which he has for many years been in the habit of frequently exploring with the miners in this low peak district of Derbyshire, he has constantly observed a deposit of mud and stony fragments similar to that which I examined with him in the Dream Cavern and the Fox Holes, and that almost all these apertures occur in elevated situations, where not a stream or rivulet exists at the present time, to the flood waters of which it would be possible to refer their introduction. The absence of pebbles and stony fragments in the interior of the cave of Fox Holes may be explained by the circumstance of there being no fissure communicating from it upwards to the surface, whilst its mouth is placed in a nearly vertical cliff that faces inwards to the valley.

In the crag called Yeo-Cliff, immediately on the west of Wirksworth, and in other natural cliffs that occur on the north side of the same town, are lofty precipices formed by the edges of the metalliferous limestone strata. These are seen to be intersected from top to bottom by numerous and nearly vertical fissures or veins, many of which are still filled with various kinds of spar and ore; others partially empty, in consequence of the removal of these substances by the operations of mining; and others again wholly or partially filled with diluvial detritus. These metalliferous limestone strata compose the upper table lands of much of this district, and have been for centuries ransacked in pursuit of lead. They are in all parts intersected by fissures similar to those exposed in the cliffs near Wirksworth, and which like them are either filled with spar, and form the most prolific deposits of metallic ore, or are choked up with mud and rocky fragments drifted in by diluvial agency, or (which is but rarely the case) remain still open, and have apparently been never filled with any thing since the fracture of the rock took place to which they owe their origin. In these last we find no other extraneous substances than a few fragments of limestone that scale off daily and fall from their sides, and the bones of cattle and recent animals that tumble in continually, and there perish.

5.—THREE SETS OF CAVES NEAR PLYMOUTH.

The fifth example I have to adduce is that of three deposits of bones discovered at Oreston, near Plymouth, by Mr. Whidby, in removing the entire mass of a hill of transition limestone for the construction of the Breakwater. The first of these is described by Sir Everard Home and Mr. Whidby, in the Philosophical Transactions for 1817. They were found in a cavern fifteen feet wide, twelve high, and forty-five long, and about four feet above high water mark; it was filled with solid clay (probably diluvial mud) in which the teeth and bones were imbedded, and was intersected in blasting away the body of the rock to make the Breakwater. The state of the teeth and bones was the same with that of those found in the caves already described; they were much broken, but not in the slightest degree rounded by attrition, and Sir Everard Home has ascertained them to belong exclusively to a species of rhinoceros. A similar discovery of teeth and bones was made in 1820, in a smaller cavern, distant one hundred and twenty yards from the former, being one foot high, eighteen wide, and twenty long, and eight feet above the high water mark; a description of its contents is given in the Philosophical Transactions for 1821, by the same gentlemen; it contained no stalactite, which abounds in many of the adjacent caverns. Sir

Everard Home describes these teeth and bones as belonging to the rhinoceros, deer, and a species of bear.

A third, and still more extensive discovery, was made in the same quarries in the summer of 1822, by the intersection of other apertures in the middle of the solid limestone, containing an immense deposit of bones and teeth imbedded in a similar earthy matrix to that in which they were found in the two former cases. On this discovery being communicated to me through the kindness of Mr. Barrow, I went immediately to Plymouth accompanied by Mr. Warburton, and found the circumstances to be nearly as follows: In a vast quarry produced by the removal of an entire hill of limestone for the construction of the Breakwater, there is an artificial cliff ninety feet in height, the face of which is perforated and intersected by large irregular cracks and cavities, which are more or less filled up with loam, sand, or stalactite. These apertures are sections of fissures and caverns that have been laid open in working away the body of the rock, and are disposed in it after the manner of chimney flues in a wall; but they attracted no attention till the discovery of bones in them last summer. Some of them have lateral communications with adjacent cavities, others are insulated and single; some rise almost vertically towards the surface, others are tortuous, passing obliquely upwards, downwards, inwards, and in all directions in the most irregular manner through the body of the rock. Apertures of the same kind occur in continual succession in the limestone of the natural cliff that forms the shore from Oreston to Stonehouse; and at the latter place, immediately on the north side of the Marine barracks, is another large quarry, in the face of which I found four apertures of the same kind leading also into caverns, the floor of all which was covered with a deep bed of mud, over the surface of which was spread a crust of stalagmite; but as these caverns have not been examined, no bones have as yet been found in them. The occurrence, however, of fissures and caves more or less filled with mud, sand, fragments of stone and stalactite, is universal in the limestone rocks of this district, whilst the dispersion of bones through them is partial. The caves of Kent's Hole and others near Babicombe and Torbay are notorious examples of this kind *.

These fissures and caverns are so connected, so often confluent and inosculating with each other, and so identical as to their contents, that there appears to be no difference as to the time or manner in which they were filled; indeed, something intermediate between a cavern and a fissure, which we may call a cavernous fissure, is the more common form under which they occur. In many of those which are nearly vertical the communication with the surface is obvious;

---

* At the extensive quarries in the rock of Chudleigh, composed of the same transition limestone as that of Plymouth, we found exactly the same phenomena of caves and fissures, partially or wholly filled with a similar diluvium, to that at Oreston. In one of the fissures so filled, which was several feet in breadth, and intersected the limestone exactly like a great wall or dyke, Mr. Warburton discovered a few small fragments of bone, and in all of them there were numerous pebbles of chert and chalk flint, and transition slate, mixed with the mud that formed the principal substance which filled these fissures up to the very surface of the soil. In another large fissure fragments of limestone predominated, and the quantity of mud was small; and here the entire mass was united by stalagmite into a breccia so solid that it remains erect, projecting like a thick wall after the removal of the limestone from each side of it, and at first sight looks more like a mass of rude masonry running across the quarry than a natural production; a similar wall, composed of fragments of oolite cemented by stalagmite, crosses the middle of a large quarry of oolite on Banner Down, near Bath. The solid rock that once inclosed it has here also been removed from both its sides, whilst at each extremity it is seen to be prolonged into the body of the natural rock, and entirely to fill up a fissure therein of considerable breadth.

whilst in others that traverse the rock obliquely it can be seen only where their upper extremity intersects the surface of the rock, and if this happens not in a cliff, but along the level face of the country, it is usually so completely filled and covered up with earth as not to be discoverable unless by approaching it through the caverns from below. We have in this circumstance a satisfactory reason why so many cavities, having at first view no apparent communication with the upper surface, are intersected in working the central and deepest portions of these limestone rocks*.

In almost all the cavities there occurs a deposit of diluvial detritus, consisting of mud and sand, and angular fragments of limestone; these substances sometimes entirely fill up the lower chambers, and are usually lodged in various quantities and proportions on the shelves and ledges, and lateral hollows of the middle and upper regions. The composition of the mud, or earthy portion of this diluvium at Plymouth, differs in some degree from that of the cave at Kirkdale, having been derived from the detritus of strata of a different character; it is of a redder colour, and looser texture, and less calculated to protect the bones in it from the access of atmospheric air and water. In one large vault at Oreston, where the quantity of diluvium is very great, it is stratified, or rather sorted and divided into laminæ of sand, earth, and clay, varying in fineness, but all referable to the diluvial washings of the adjacent country. It is also partially interspersed with small fragments of clay-slate and quartz. The sand and loam are in many places invested with, and cemented together by, stalagmite, but

---

* See a good illustration of this in Plate XX., and the description of the cave at the Dream Mine, near Wirksworth, at page 62, et seq.

not so firmly as in the Gibraltar breccia, and in much of that which occurs in the caves of Germany : portions of bone and single fragments of rock are also found occasionally incased with a thin crust or coating of the same substance ; but, generally speaking, it is not sufficiently abundant to hold the mud and bones together in a solid mass after they are moved from the cave. In some few spots there were balls of iron stone, and concretions of ochre formed in the clay ; in others there was a considerable deposit of manganese ore dispersed through the sand and porous portions of the loam ; in the latter were also concentric balls of the same ore inclosed within each other after the manner of the ochreous aetites *.

It was in one of these oblique apertures in the present face of the rock at Oreston, and at about forty feet above the bottom of the quarry, that the congeries of bones, skulls, horns, and teeth I am now about to speak of, was discovered last summer. Mr. Whitby had collected fifteen large maund baskets full of them before our arrival ; these have been sent to the College of Surgeons, and distributed to various public collections. In the upper parts of the cavity from

* A small quantity of manganese is found disposed in a similar manner in some parts of the sediment in the cave at Kirkdale ; and I have often found this ore incrusting the pebbles and sand in subordinate beds of a large mass of mixed diluvial gravel, e g. in Lord Barrington's gravel pit at Sedgefield, in the county of Durham, and near Namur. It does not pervade the entire mass, but is usually limited to beds of a few inches in thickness, and seldom spreads in these beyond a few feet in breadth. The source from which it has been derived in situations of this kind is by no means obvious ; it is decidedly of later origin than the gravel itself, and seems analogous to the incrustations of iron and carbonate of lime, which are often disposed in a similar manner around the component materials of common surface gravel, cementing it into a solid breccia, (e. g. in the gravel pits immediately on the east of Oxford). It seems to have been deposited from water infiltrated through the gravel and sand which are thus incrusted by it.

which they were taken we saw appearances of as many more still undisturbed, and forming a mass which entirely blocked it up to an extent which we could not then ascertain ; those already extracted had been discovered in a hollow, which apparently formed the lowest part of a cavernous aperture descending obliquely downwards.   Ascending this aperture from its lowest point, we pursued it upwards many feet through the solid rock till it became narrow, and was entirely obstructed by a mass of bones, fragments of limestone, and mud ; these have probably been since removed, and I must refer to Mr. Whitby's account, now before the Royal Society, for further particulars as to the manner in which he has traced its connexions upwards through other caverns towards the surface *.

* In the course of the last summer Joseph Cottle, Esq. of Bristol, has made a large collection of bones from this same cave during a visit to Plymouth. He has added the tiger to the list of animals before discovered in it.   He has favoured me with the following list of the remains in his possession.

    18 jaws of horse.
     2 jaws of ox.
     2 jaws of hyæna.
     2 jaws of deer.
     5 jaws of wolf.
Single teeth, 188 of horse.
         26 of ox.
          9 of hyæna.
         2 tusks of tiger: one $3\frac{3}{4}$ inches long, the other $3\frac{1}{4}$; one from the upper, the other from the lower jaw.
         5 teeth of wolf.
        35 of deer.
        50 of ox or deer: not ascertained.

Bones: 300 large and small, chiefly of the horse; none of them are gnawed, many are quite perfect, and the majority of them slightly broken.

Osseous breccia: 33 specimens, containing teeth and bones cemented by stalagmite.

From this list it appears that the bones of horse greatly predominate in the collec-

The bones appeared to us to have been washed down from above at the same time with the mud and fragments of limestone, through which they are dispersed, and to have been lodged wherever there was a ledge or cavity sufficiently capacious to receive them, or a strait sufficiently narrow to be completely obstructed by them; they were entirely without order, and not in entire skeletons; occasionally fractured, but not rolled; apparently drifted, but to a short distance from the spot in which the animals died; they seem to agree in all their circumstances with the osseous breccia of Gibraltar, excepting the accident of their being less firmly cemented by stalagmitic infiltrations through their earthy matrix, and consequently being more decayed; they do not appear, like those at Kirkdale, to bear marks of having been gnawed or fractured by the teeth of hyænas, nor is there any reason to believe them to have been introduced by the agency of these animals.

The only marks I have seen on them were those pointed out to me by Mr. Cliff, of nibbling by the incisor and canine teeth of an animal of the size of a weasel, showing distinctly the different effect of each individual tooth on the ulna of a wolf, and the tibia of a horse; and a few pits or circular cavities produced by partial decomposition on one surface only of the tibia of an ox, exactly resembling those which occur on many of the bones from the cave at Kirkdale.

These pits must have been formed before the bone was imbedded

tion made by Mr. Cottle: in that sent to the College of Surgeons those of the ox were much more numerous, being nearly equal to those of the horse; but whatever be the disproportion of their numbers, the bones and teeth of all the animals are found confusedly mixed together in irregular heaps, and not in entire skeletons, nor arranged in different parts of the cavern according to the difference of their species.

L

in mud in the lowest recesses of the cave, and probably whilst it lay exposed in some upper cavity of the rock.   The weasels' teeth also must have made their impressions on the bones of the wolf and horse before they were buried in diluvial mud, and probably whilst these dead animals lay in the same situation with the tibia of the ox.

The bones when half dry, on being thrown out of a basket on the floor, had the smell of a charnel-house, or newly opened grave.   On examining the spot where they lay yet undisturbed in the mud of the cave, we found some of them decomposed, and crumbling under the touch into a blackish powder, and all extremely tender and fran- gible, and of a dark brown colour whilst wet; but on drying they acquired a greater degree of firmness and a whiter colour.   They retain less of their animal gelatin than the bones at Kirkdale, and when dry they ring if a blow be given to them, and are absorbent to the tongue.   On some of them there are marks of extensive disease.

Mr. Cliff has discovered in two of these bones from Oreston (the metatarsus and metacarpus of an ox,) extensive enlargement by ossific inflammation, arising probably from a kick or blow; and also cavities and swellings produced by abscesses in both sides of the under jaw of a wolf.   Professor Soemmering also has in his collection the head of an hyæna from the cave of Gailenreuth, from which part of the nose, with the canine and incisor teeth, had been entirely torn away, and the elevated ridge formed by the parietal crest and frontal suture dreadfully lacerated by the fangs of a bear, or tiger, or some more powerful animal, and the individual had survived until the in- juries had been considerably, and, as far as was possible, repaired.   A drawing of this head will be given in the 4th vol. of the 2d edition of Cuvier's Animaux Fossiles.   M. Esper, also, in his Representations of bones from Gailenreuth (Plate XIV.), gives a drawing of the thigh bone of a bear, which had been broken and re-united, having at the points of junction marks of extensive exostosis.

It appears from a description accompanied by beautiful drawings of many of the Oreston bones by Mr. Cliff, which will be published in the Phil. Trans. for 1823, that those sent to the College of Surgeons belong to the six following genera of animals, viz. Hyæna, Wolf, Fox, Horse, Ox, and Deer; to these must be added the tusks of Tiger, discovered by Mr. Cottle.   Mr. Cliff has ascertained the following number of individuals.

Hyænas, five or six, of the same extinct species as those at Kirkdale; three of them were young, and in the act of changing their teeth.

Wolves, five.   Mr. Cliff can find no difference between these fossil teeth and bones and those of existing species*.

Fox, two tusks, much decayed, and absorbent to the tongue.

Horses, about twelve, of different ages and sizes, as if from more than one species.

Oxen, about twelve, of different species, some having very short and straight horns; but not referable to young animals.

Deer, two or three, of a small species.

No traces of bear or rhinoceros have yet been noticed in this last discovered cavern, though they occurred in the former cavities at no great distance; but this circumstance is an accident likely to arise

---

* M. Cuvier also can find no difference between the bones and teeth of fossil wolves and foxes, which he has examined, and those of recent ones.

from the irregular manner in which the remains are now dispersed, and implies no difference in the time or circumstances under which they were introduced.

It remains only to consider what this time and what the circumstances were. I have already stated, that there is no evidence like that at Kirkdale to show the animal remains at Oreston to have been collected by the hyænas; no disproportion in the number of the teeth to that of the bones; no destruction of the condyles and softer parts, and abundance in excess of fragments of the harder portions; no splinters of the marrow bones; no friction or polish on the convex surfaces only of the curved bones; no marks of large teeth; no album græcum; and no dispersion of bones along the horizontal surface of a habitable den: but, on the contrary, a deep hole nearly perpendicular, and bones quite perfect, lodged in irregular heaps in the lowest pits, and in cavities along the lateral enlargements of this hole, and mixed with mud, pebbles, and fragments of limestone, in precisely the same manner as I shall hereafter show them to be lodged and mixed in the caves and fissures of Germany and Gibraltar; and as they would have been, supposing they were drifted to their present place by the diluvian waters from some lodgment which they had before obtained in the upper regions of these extensive and connected cavities. That they are of antediluvian origin is evident from the presence of the extinct hyæna, tiger, and rhinoceros; but there still remains a difficulty in ascertaining what was the place from which they were so drifted; 1. Are they the bones of animals that were drowned, and their bodies drifted in entire by the waters which introduced the mud and pebbles? Or, 2, had they laid some time dead on the antediluvian surface of the earth, till they were washed in at the deluge? Or, 3, were they derived from the animals that had fallen into the open antediluvian fissures, and there perishing, remained as entire skeletons in the spots on which they died, till they were drifted on further by the diluvian waters into the lowest recesses and under-vaultings with which these fissures had communication, and there mixed up, in irregular heaps, with mud, pebbles, and angular fragments of limestone, all falling down together with them to the places of their present interment, and producing in this short transit that quantity of fracture to which they have been submitted?

1. On the first of these hypotheses, had they been drowned, and the carcases drifted in by the diluvian waters, we should have found the skeletons more entire, and the bones less broken and less confusedly mixed together than they are; and we should neither have had the marks of nibbling by the weasel's teeth on the bones of the wolf and horse, nor the hollow pits arising from partial decay on one surface only of the tibia of the ox; for neither of these effects could have been produced on bones surrounded with a bed of mud.

2. To the second hypothesis, that they had laid as dead bones on the antediluvian surface till they were drifted from thence into the fissures, I would reply, that in a land inhabited as this was by wolves and hyænas, it is not likely that any carcases would have laid long on the surface without at least the softer portions of the bones being eaten off by the hyænas, and thus we should have found them lacerated rather than perfect in the place to which they have since been drifted;

they might also in this case have been expected to be more or less rolled, and to have lost their angles by friction, which does not appear to be the fact. Another objection also arises from this circumstance, that the bones of dead animals exposed on the surface of the earth, without any protection of soil or gravel, are soon destroyed by minute insects and continual atmospheric changes; and were it not so, the world would by this time have been spread over most abundantly with the bones of the myriads of animals that have died on its surface, and received no burial ever since the period of the last retreat of the diluvial waters.

3. The third hypothesis is that which I propose as most probable, viz. that the animals had fallen during the antediluvian period into the open fissures, and there perishing, had remained undisturbed in the spot on which they died, till drifted forwards by the diluvian waters to their present place in the lowest vaultings with which these fissures had communication. This explanation is supported by the strong fact, that animals at this day do fall continually into the few fissures that are still open, and that carnivorous as well as grami-nivorous animals lie in nearly entire skeletons in the open fissure at Duncombe Park, each in the spot on which it actually perished, upon the different ledges and landing places that occur in the course of its descent, and from which, if a second deluge were admitted to this fissure, it could only drift them downwards, and with them the loose angular fragments amidst which they now lie, to the lowest chambers in which the bottom of this fissure terminates. The teeth marks of the weasel, and the pitted surface of the tibia, will on this hypothesis

have been effects produced on the bones as they lay dead within the fissures (for a weasel might find access by minute crevices to the interior of such fissures), and the wolves and hyænas may have either fallen, like the horses, oxen, and deer, by accident into these natural pitfalls, or have been tempted to the fatal experiment of leaping into them by the carcases of the other animals, whilst they lay yet undecayed within the fissures. The proportion of individuals collected at Oreston (the graminivorous being very much in excess beyond the carnivorous) is, as far as it goes, consistent with this hypothesis; and if this solution appears fanciful, it is one that need not be urged, for by the same accident that dogs at. this day fall into the open fissure at Duncombe Park, no less than sheep and deer, might the wolves and the hyænas also of the antediluvian world have fallen, as well as the horses and oxen, into the chasms which then in countless numbers crossed their paths, whenever they ventured on the perilous regions of the hollow and fissile limestone; and possibly some of them, whilst in the very act of pursuing their prey, may have dashed (like our less ferocious dogs in pursuit of game) into the chasms, which became the common grave of themselves and of the victim they were too eager to devour. And however new and unheard-of the existence of such fissures may be to those who have never visited or lived in a country composed of compact limestone, it is matter of painful notoriety to the farmers in Derbyshire, that their cattle are often lost by falling into the still open fissures that traverse the districts of the Peak; and it is no less matter of fact, that similar accidents are avoided in the mountain limestone countries of Monmouth and Glamorganshire only by walls carefully erected round all the open chasms, with which there also the same rocks are intersected*.

In speaking of the bones at Oreston in my former paper on Kirkdale, I had expressed a decided opinion that the caverns in which they occur must have had some communication with the surface through which the bones may have been introduced; and as Mr. Whidby has since found reason to adopt the same opinion in a further account of this third discovery to the Royal Society, accompanied by plans and sections of the caves, and Mr. Cliff has laid before the same society an anatomical description of the bones, with beautiful drawings, all of which will probably appear in the Phil. Trans. for 1823, I shall conclude this part of my subject with referring my readers to these memoirs for further particulars.

6.—CAVE OF CRAWLEY ROCKS, NEAR SWANSEA.

The sixth deposit of bones which has come to my knowledge was in the parish of Nicholaston, on the coast of Glamorganshire, at a spot called Crawley Rocks, in Oxwich Bay, about twelve miles S.W. of Swansea; it was discovered in the year 1792, in a quarry of limestone, on the property of T. M. Talbot, Esq. of Penrice Castle, and no account of it has, I believe, been ever published; some of the bones

---

* In Sir John Nicholl's park at Merthyr Mawr there are many such apertures thus walled round; and in mining countries we know that animals are perpetually being lost by falling into old shafts that are not sufficiently fenced round to keep them off.

however are preserved in the collection of Miss Talbot, at Penrice; they are as follows:

Elephant.......Three portions of large molar teeth.

Rhinoceros.....Right and left ossa humeri.

One atlas bone.

Two molar teeth of upper jaw.

Ox..............First phalangal bone of left fore foot.

Stag............Lower extremity of the horn.

Three molar teeth.

One first phalangal bone, right leg.

Hyæna.........Two canine teeth, much worn.

These bones were found in a cavity of mountain limestone, which was accidentally intersected, like the cave at Kirkdale, in working a quarry: they have a slight ochreous incrustation, and a little earthy matter adhering to them; but are not in the least degree rolled; and the condyles of the two humeri of the rhinoceros, belonging to different individuals, have in each case been entirely broken off. There is also in the collection of J. Lucas, Esq. at Southall, in this neighbourhood, the entire femur of a rhinoceros, said to have been found many years ago in a cavern of limestone at Port Inon, together with teeth, and a gigantic skull, which was sent over to Appledore, and has not been heard of since. As there is a similar tradition of a large skull having been found at Crawley Rocks, together with the bones now at Penrice, it is probable that this head, and possibly the femur of the rhinoceros also, were found all together in the cave at Crawley, which has now been entirely cut away.

M

7.—CAVE OF PAVILAND.

The seventh and last case that has occurred in this country is that of another discovery recently made on the coast of Glamorganshire, fifteen miles west of Swansea, between Oxwich Bay and the Worms Head, on the property of Earl Talbot. It consists of two large caves facing the sea, in the front of a lofty cliff of limestone, which rises more than 100 feet perpendicularly above the mouth of the caves, and below them slopes at an angle of about 40° to the water's edge, presenting a bluff and rugged shore to the waves, which are very violent along this north coast of the estuary of the Severn. These caves are altogether invisible from the land side, and are accessible only at low water, except by dangerous climbing along the face of a nearly precipitous cliff, composed entirely of compact mountain limestone, which dips north at an angle of about 45°. One of them only (called Goats Hole) had been noticed when I arrived there, and I shall describe it first, before I proceed to speak of the other. Its existence had been long known to the farmers of the adjacent lands, as well as the fact of its containing large bones, but it had been no farther attended to till last summer, when it was explored by the surgeon and curate of the nearest village, Port Inon, who discovered in it two molar teeth of elephant, and a portion of a large curved tusk, which latter they buried again in the earth, where it remained till it was extracted again, on a further examination of the cave in the end of December last by L. W. Dillwyn, Esq. and Miss Talbot, and removed to Penrice Castle, together with a large part of the skull to which it had belonged, and several baskets full of other teeth and bones. On the news of this further discovery being communicated to me, I went immediately from Derbyshire to Wales, and found the position of the cave to be such as I have above described; and its floor at the mouth to be from 30 to 40 feet above high-water mark, so that the waves of the highest storms occasionally dash into it, and have produced three or four deep rock basins in its very threshold, by the rolling on their axis of large stones, which still lie at the bottom of these basins (see Plate XXI. H H.); around their edge, and in the outer part of the cave itself, are strewed a considerable number of sea pebbles, resting on the native limestone rock. The floor of the cave ascends rapidly from its mouth inwards to the furthest extremity (see Plate XXI. and description), so that the pebbles have not been drifted in beyond twenty feet, or about one-third of its whole length; in the remaining two-thirds no disturbance by the waters of the present sea appears ever to have taken place, and within this point at which the pebbles cease, the floor is covered with a mass of diluvial loam of a reddish yellow colour, abundantly mixed with angular fragments of limestone and broken calcareous spar, and interspersed with recent sea-shells, and with teeth and bones of the following animals, viz. elephant, rhinoceros, bear, hyæna, wolf, fox, horse, ox, deer of two or three species, water-rats, sheep, birds, and men. I found also fragments of charcoal, and a small flint, the edges of which had been chipped off, as if by striking a light. I subjoin a list of the most remarkable of the animal remains, most of which are preserved in the collection at Penrice Castle.

M 2

Elephant......Head broken into numerous fragments, the sockets of
the tusks being nearly entire, and six inches in
diameter.

One large portion of tusk, nearly two feet long, and five
inches and a half in diameter.

One large portion of diseased tusk, and many very small
fragments of decayed ivory.

Two molar teeth entire, fragments of two others.

Part of the epiphysis of the humerus.

Large fragments of the ribs.

Splinters of large cylindrical bones of the legs.

Rhinoceros ....A tooth resembling the incisor of the upper jaw.

One fragment of upper molar tooth.

Two phalangal bones of the toe.

Horse ....Many teeth and fragments of bones.

Hog ....One upper incisor, apparently modern.

Bear...........Many molar teeth, two canine ditto.

One fragment of lower jaw, and the anterior portion or
chin part of two other lower jaws firmly anchelosed,
and exhibiting the sockets of the incisor teeth and
of both tusks; the latter are more than three inches
deep, and equal in size to the largest from the caves
of Germany.

One humerus, of the same large size, nearly entire.

Many vertebræ, equally large.

Two ossa calcis, and many large bones of the metacarpus and metatarsus.

Hyæna.........Lower extremity of the left humerus.

Fox.............Lower extremity of the femur.

Wolf...........One lower jaw.

One os calcis.

Several metacarpal bones.

Ox..............Many teeth.

Two lumbar vertebræ.

One femur, and many entire bones of the foot, and fragments of larger bones.

Deer.........One skull, large as the red deer, but of a different species.

Fragments of various horns, some small, others a little palmated, one approaching to that of the roe.

Many teeth, and fragments of bones.

Rat..........One skeleton, nearly entire, of a small water-rat, or large field-mouse, probably postdiluvian.

Birds........Single bones of small birds, all recent.

Man.........Portion of a female skeleton, clearly postdiluvian.

Fragments of many recent bones of ox and sheep, apparently the remains of human food.

The entire mass through which the bones are dispersed appears to have been disturbed by ancient diggings, and its antediluvian remains thereby to have become mixed with recent bones and shells; the latter of which Mr. Dillwyn has examined, and refers to the following species: buccinum undatum, turbo littoreous, patella vulgata, trochus crassus, nerita littoralis; these are all common on the adjacent shore, and the animals that inhabit them are all eatable. That portion of the diluvial mass which lies on the east side of the cave (see Plate

XXI. F.) adheres together in a loose breccia, and has been less disturbed than the rest, which it overhangs with a cliff about five feet high, and extending inwards from F to the interior extremity of the cave B, where it enters into and covers the floor of the small hole that terminates the cave. At the point B the recent shells and bones of birds are most abundant, and the earthy mass containing them is cemented to a firm breccia by stalagmite; and this is almost the only point within the cave at which any stalagmite or stalactite occurs. The two elephants' teeth were found in the small cliff F, at a distance from the head and tusk, which lay close together in the loose earth E, at the spot represented in the drawing. The anterior part of the skull, and the sockets of both the tusks, were found nearly entire, but have been much broken by removal. They were but slightly covered with earth, and very tender; the portion of tusk also, being about two feet long, is so much decayed that the whole of its interior has crumbled to small angular fragments, so soft as to be cut by the nail, whilst the outer laminæ alone remain entire, and in the form of a hollow shell, which is preserved at Penrice; so also are the fragments that composed great part of the entire skull, and were broken in extracting them; and another portion of ivory, in which has been formed an irregular cavity, about two inches in diameter, similar to those effects of ossific inflammation which are produced in recent ivory by gun-shot wounds, and encircled with concentric laminæ of bony matter, placed obliquely to the grain of the ivory: it is probably the effect of a blow or puncture received whilst this part of the tusk was yet in its pulpy state, and within the socket. No large bones of the skeleton have as yet been discovered entire; they seem to have been destroyed and broken to pieces by repeated diggings. The other ancient bones also have been much broken, and appear generally in the state of fragments dispersed irregularly through the earthy matrix, together with ancient teeth, and fragments of horn, and with the modern bones and recent shells above enumerated. None of these remains have any marks of having been gnawed or rolled, nor have the fragments of limestone, and of calcareous spar that occur with them, lost much of their angles. Among the horns I noticed the base of two that are separate from the skull, and appear to have been cast off by necrosis; and among the bones was the entire skull of a deer, from which the horns had been broken off by violence. In the centre of the cave, and about two feet deep, I found under and amongst the broken bones of elephant, bear, and other extinct animals, a portion of the scapula apparently of a sheep, which had been smoothly cut across as if by a butcher's saw; and, from its state of preservation, was decidedly not antediluvian. This mixture of ancient and comparatively modern bones must have arisen from repeated diggings in the bottom of the cave.

In another part (see Plate XXI.) I discovered beneath a shallow covering of six inches of earth nearly the entire left side of a human female skeleton. The skull and vertebræ, and extremities of the right side were wanting; the remaining parts lay extended in the usual position of burial, and in their natural order of contact, and consisted of the humerus, radius, and ulna of the left arm, the hand being wanting; the left leg and foot entire to the extremity of the toes, part of the right foot, the pelvis, and many ribs; in the middle of the bones of the ancle was a small quantity of yellow wax-like sub-

stance resembling adipocere.   All these bones appeared not to have
been disturbed by the previous operations (whatever they were) that
had removed the other parts of the skeleton.   They were all of them
stained superficially with a dark brick-red colour, and enveloped by a
coating of a kind of ruddle, composed of red micaceous oxyde of iron,
which stained the earth, and in some parts extended itself to the
distance of about half an inch around the surface of the bones.   The
body must have been entirely surrounded or covered over at the
time of its interment with this red substance.   Close to that part of
the thigh bone where the pocket is usually worn, I found laid together,
and surrounded also by ruddle, about two handsfull of small shells
of the nerita littoralis in a state of complete decay, and falling to
dust on the slightest pressure.   At another part of the skeleton, viz.
in contact with the ribs, I found forty or fifty fragments of small
ivory rods nearly cylindrical, and varying in diameter from a quarter
to three quarters of an inch, and from one to four inches in length.
Their external surface was smooth in a few which were least decayed; but the greater number had undergone the same degree of
decomposition with the large fragments of tusk before mentioned;
most of them were also split transversely by recent fracture in digging
them out, so that there are no means of knowing what was their
original length, as I found none in which both extremities were unbroken; many of them also are split longitudinally by the separation
of their laminæ, which are evidently the laminæ of the large tusk,
from a portion of which they have been made.   The surfaces exposed
by this splitting, as well as the outer circumference where it was
smooth, were covered with small clusters of minute and extremely

delicate dendrites; so also was the circumference of some small fragments of rings made of the same ivory, and found with the rods,
being nearly of the size and shape of segments of a small teacup
handle; the rings when complete were probably four or five inches
in diameter.   Both rods and rings, as well as the nerite shells, were
stained superficially with red, and lay in the same red substance that
enveloped the bones; they had evidently been buried at the same
time with the woman.   In another place were found three fragments
of the same ivory, which had been cut into unmeaning forms by a
rough edged instrument, probably a coarse knife, the marks of which
remain on all their surfaces.   One of these fragments is nearly of the
shape and size of a human tongue, and its surface is smooth as if it
had been applied to some use in which it became polished, and by
which the scratches of the coarse knife from which it received its
shape have been nearly obliterated: there was found also a rude instrument, resembling a short skewer or chopstick, and made of the
metacarpal bone of a wolf, sharp and flattened to an edge at one end,
and terminated at the other by the natural rounded condyle of the
bone, which the person who cut it had probably extracted, as well as
the ivory tusk, from the diluvial detritus within the cave.   No metallic instruments have as yet been discovered amongst these remains,
which, though clearly not coeval with the antediluvian bones of the
extinct species, appear to have lain there many centuries.

The charcoal and fragments of recent bone that are apparently
the remains of human food, render it probable that this exposed and
solitary cave has at some time or other been the scene of human habitation, if to no other persons, at least to the woman whose bones I

N

have been describing.   The ivory rods and rings, and tongue-shaped fragments, are certainly made from part of the antediluvian tusks that lay in the same cave; and as they must have been cut to their present shape at a time when the ivory was hard, and not crumbling to pieces as it is at present on the slightest touch, we may from this circumstance assume to them a very high antiquity, which is further confirmed by the decayed state of the shells that lay in contact with the thigh bone, and, like the rods and rings, must have been buried with the woman.   The wolf's toe bone also was probably reduced to its present form, and used by her as a skewer, the immediate neighbourhood being wholly destitute of wood.

The circumstance of the remains of a British camp existing on the hill immediately above this cave, seems to throw much light on the character and date of the woman under consideration; and whatever may have been her occupation, the vicinity of a camp would afford a motive for residence, as well as the means of subsistence, in what is now so exposed and uninviting a solitude.   The fragments of charcoal, and recent bones of oxen, sheep, and pigs, are probably the remains of culinary operations; the larger shells may have been collected also for food from the adjacent shore, and the small nerite shells either have been kept in the pocket for the beauty of their yellow colour, or have been used, as I am informed by the Rev. Henry Knight, of Newton Nottage, they now are in that part of Glamorganshire, in some simple species of game.   The ivory rods also may have either been applicable to some game, as we use chess men or pins on a cribbage board; or they may be fragments of pins, such as Sir Richard Hoare has found in the barrows of Wilts and

Dorset, together with large bodkins also of ivory, and which were probably used to fasten together the coarse garments of the ancient Britons. It is a curious coincidence also, that he has found in a barrow near Warminster, at Cop Head Hill, the shell of a nerite, and some ivory beads, which were laid by the skeletons of an infant and an adult female, apparently its mother *.

That ivory rings were at that time used as armlets, is probable from the circumstance of similar rings having also been found by Sir Richard Hoare in these same barrows; and from a passage in Strabo, lib. 4, which Mr. Knight has pointed out to me, in which, speaking of the small taxes which it was possible to levy on the Britons, he specifies their imports to be very insignificant, consisting chiefly of ivory armlets and necklaces, Ligurian stones, glass vessels, and other such like trifles. The custom of burying with their possessors the ornaments and chief utensils of the deceased, is evident from the remains of this kind discovered every where in the ancient barrows; and this may explain the circumstance of our finding with the bones of the woman at Paviland the ivory rods, and rings, and nerite shells, which she had probably made use of during life. I am at a loss to conjecture what could have been the object of collecting the red oxyde of iron that seems to have been thrown over the body when

---

* A long and rude shaped pin made of bone, of very high antiquity, being of the size and length of a large wooden skewer, and very similar to the smaller fragments of ivory from Paviland, has recently been found on Foxcomb hill, near Oxford; and my friend the Rev. J. J. Conybeare has discovered a bone bodkin, nearly of the same size, among the remains of the British or Belgic settlements which he has lately been tracing out with great success on the flat summits called Charmy Down, Banner Down, Salisbury, and Claverton Down, in the immediate neighbourhood of Bath.

laid in the grave : it is a substance, however, which occurs abundantly in the limestone rocks of the neighbourhood.

The disturbed state of the diluvial earth all over the bottom of the cave, and fractured condition of the ancient bones, may have been produced by digging in search of more ivory, or to gratify the curiosity which the discovery of such large and numerous remains must naturally have excited ; and in the course of these diggings the antediluvian bones would become mixed with those of modern animals which had been introduced for food. The preservation of so large a part of the elephant's tusk may probably have arisen from the use to which it was destined, and had been in part appropriated in the making of rods and rings.

From all these circumstances there is reason to conclude, that the date of these human bones is coeval with that of the military occupation of the adjacent summits, and anterior to, or coeval with, the Roman invasion of this country.

The above are the most remarkable phenomena in the interior of this cave. It remains only to describe a long cavernous aperture that rises like a crooked chimney from its roof to the nearly vertical face of the rock above : its form and diameter are throughout irregular, the latter being about twelve feet where longest, and in its narrowest part about three feet ; so that it is impossible the large elephant, whose bones were found in the cave below, could have been drifted down entire through this aperture. It expands and contracts irregularly from D, its lower extremity in the roof of the cavern, to K, the point at which it terminates in the face of the cliff. (See Plate XXI.) Along this tortuous ascent are several lateral cavities, L. L. L., the bottoms of which afford a place of lodgment for a bed of brown earth about a foot thick, and derived apparently from dust driven in by the wind. In this earth I found the bones of various birds, of moles, water-rats, mice, and fish, and a few land shells; all these are clearly the remains of modern animals, and their presence in this almost inaccessible spot can only be explained by referring the bones of birds, moles, rats, and mice, to the agency of hawks, and the fish-bones to that of sea-gulls. The land shells are such as live at present on the rock without, and may easily have fallen in. Had there been any stalagmite uniting these bones into a breccia, they would have afforded a perfect analogy to the accumulation of modern birds' bones, by the agency of hawks, at Gibraltar; where Major Imrie describes them as forming a breccia of modern origin in fissures of the same rock which has other cavities filled with a bony breccia of more ancient date, and which I shall presently endeavour to show is of the same antediluvian origin with the older parts of the bones that occur on the floor of the cave at Paviland.

Whilst exploring this cavern, I was informed by the workmen that there was another of the same kind about a hundred yards further to the west; and proceeding to examine it, I found it to be very similar to the first, in size, form, and position, and closed on every side with solid rock, excepting the mouth, which is large and open to the sea; its body contracts gradually towards the inner extremity, and upwards also towards the roof, where it terminates in a vein, that is still filled with calcareous spar: the cave itself, in fact, seems to be merely an enlargement of this vein. There is also a similar, but longer and more narrow, aperture immediately on the east of Goats Hole, the bottom of which, being on the level of the sea, is almost perpetually under water. This east cave also is seen to terminate upwards in a vein of calcareous spar. The floor of the west cave is at its mouth about thirty feet above the sea, and more horizontal than that of Goats Hole, and being throughout within reach of the highest storm waves, is strewed over entirely, to the depth of more than a foot, with a bed of small sea pebbles. Digging through these, I found beneath them a bed of the same argillaceous loam and fragments of limestone as in the Goats Hole, and a still more abundant accumulation of animal remains. In a short time I collected two baskets' full of the teeth and bones of ox, horse, deer, and bear; and have reason to think the entire floor beneath the pebbles is covered with a continuous mass of the same diluvial earth and fragments of stones, intermixed with teeth and bones, and altogether of the same age and origin with the antediluvian part of those in Goats Hole, the near position of which renders it probable that both these caves are residuary offshoots or branches of some larger cavern, that has been cut away by the denudation which formed the present cliffs, and whose main trunk is now no more; and that by means of this main trunk they originally had communication with each other, and received at the same time the animal remains and diluvial detritus that are common to them both. Their relative position is such, that if both were prolonged towards the sea they would soon meet, and either become confluent, or intersect each other.

The time and manner in which these two caverns received the antediluvian teeth and bones, and the earthy matter through which they are dispersed, would not so easily have been ascertained had it not been in our power to illustrate them by the analogies of other caverns now under consideration.   From a comparison of these with the internal evidence afforded at Paviland, it seems nearly certain that the latter are identical in all the circumstances of their diluvial and antediluvial phenomena with those of the former; and that occurring as they do, in the vertical cliffs that flank the submarine valley which forms the estuary of the Severn, they are analogous to the caves we find in the equally vertical and not less lofty cliffs that flank the inland valleys of the Avon at Clifton, of the Weissent river at Muggendorf, of the Bode river at Rubeland in the Hartz, and of the Mur at Peckaw, near Gratz, in Styria; all being cliffs produced by diluvial denudation, and all containing, in a nearly vertical precipice, the mouths of caves which are but the truncated extremities of other and originally more extensive caverns, which descended from the antediluvian surface, and terminated in the vaults that still remain in those portions of the rock which have not been washed away by the diluvial waters, from whose action these cliffs have derived their origin.   By such larger and upper chambers, whose destruction I am now assuming (and for the proof of which I must refer to the concluding part of this work), the animal remains may either have been washed in at the same time with the diluvial loam and fragments of stone, in the midst of which they lie, or have fallen in and perished in the period immediately preceding the deluge, and been subsequently drifted onwards to their present place in the lowest recesses with which the upper cavities had communication.   The detail of the manner in which this latter process may have taken place has been already pointed out in my description of the caves at Oreston, near Plymouth.   I have as yet found no evidence to show that either of the caves at Paviland were occupied as antediluvian dens.

In the flat surface of the fields, a quarter of a mile distant inland from the cliff of Paviland, is an open cavern, to which it is possible to descend only by a ladder, and which, like the open fissure at Duncombe Park, contains at its bottom, and in the course of its descent, the uncovered skeletons of sheep, dogs, foxes, and other modern animals, that occasionally fall into it and perish.   It is needless to repeat the arguments I have founded on facts of this kind, to show the manner in which antediluvian animals may have fallen into the then existing cavities of the limestone rocks, and have supplied the remains we find in the bony breccia of Gibraltar and Plymouth, and I may here add also of Paviland.

The above facts are, I think, sufficient to warrant us in concluding, that in the period we have been speaking of the extinct species of hyæna, tiger, bear, elephant, rhinoceros, and hippopotamus, no less than the wolves, foxes, horses, oxen, deer, and other animals which are not distinguishable from existing species, had established themselves from one extremity of England to the other, from the caves of Yorkshire to those of Plymouth and Glamorganshire; whilst the diluvial gravel beds of Warwickshire, Oxford, and London, show that they were not wanting also in the more central parts of the country; and M. Cuvier has established, on evidence of a similar nature, the probability of their having been spread in equal abundance over the Continent of Europe.   But it by no means follows, from the certainty of the bones having been dragged by beasts of prey into the small cavern at Kirkdale, that those of similar animals must have been introduced in all other cases in the same manner; for as all these animals were the antediluvian inhabitants of the countries in which the caves occur, it is possible, that some may have retired into them to die, others have fallen into the fissures by accident and there perished, and others have been washed in by the diluvial waters. By some one or more of these three latter hypotheses, we may explain those cases in which the bones are few in number and not gnawed, the caverns large, and the fissures extending upwards to the surface; but where they bear marks of having been lacerated by beasts of prey, and where the cavern is small, and the number of bones and teeth so great, and so disproportionate to each other as in the cave at Kirkdale, the only adequate explanation is, that they were collected by the agency of wild beasts. We shall show hereafter, that in the case of the German caves, where the quantity of bones is greater than could have been supplied by ten times the number of carcases which the caves, if crammed to the full, could ever have contained at one time, they were derived from bears that lived and died in them during successive generations.

Although it must appear probable from the facts I have now advanced, that similar bones abound generally in the caves and fissures of our limestone districts, we shall yet cease to wonder that their existence has been so long unnoticed, when we consider the number of accidental circumstances that must concur to make them objects of public attention. 1st, The existence of caverns is an accidental occurrence in the interior of the rock, of which the external surface affords no indication, when the mouth is filled with rubbish, and overgrown with grass, as it usually is in all places, excepting

o

cliffs and the face of stone quarries. 2d, The presence of bones is another accident, though probably not an uncommon one in those cavities which were accessible to the wild animals, either falling in, or entering spontaneously, or being dragged in by beasts of prey, in the period immediately preceding the deluge. 3d, A further requisite is, the intersection of one of these cavities, in which there happen to be bones, by a third accident, viz. the working of a stone quarry, by men who happen to have sufficient curiosity or intelligence to notice and speak of what they find, and this to persons who also happen to be willing or able to appreciate and give publicity to the discovery. The necessary concurrence of all these complicated contingencies renders it probable, that however great may be the number of subterraneous caverns, in an inland country, very few of them will ever be discovered, or if discovered, be duly appreciated. Those I have mentioned in Yorkshire, Devon, Somerset, Derby, and Glamorganshire, were all laid open, with the exception of the caves at Paviland, by the accidental operations of a quarry or mine.

CAVES OF GERMANY.

WE may now proceed to consider how far the circumstances of the caves we have been examining in England appear consistent with those of analogous caverns in other parts of the world. The history of the diluvian gravel of the Continent, and of the animal remains contained in it, appears altogether identical with that of those in our own country; and with respect to the bones that occur in caverns, the chief difference seems to be, that on the Continent some of the caves have their mouths open, and have been inhabited also in the post-diluvian period by animals of existing species. Thus at Gailenreuth the great extinct bear (Ursus spelæus) occurs, together with the Yorkshire species of extinct hyæna, in a cave, the mouth of which has no appearance of having ever been closed, and which at this moment would, probably, have been tenanted by wild beasts, had not the progress of human population extirpated them from that part of Germany.

For the best existing accounts of the cavern at Gailenreuth, which I have twice visited in 1816 and 1822, and of which, in Plate XVII. I have given a sectional representation, I must refer to Esper's "Description des Zoolithes et des Cavernes dans le Margraviat de Bareuth," fol. 1774, with fourteen plates of the bones of bears and hyænas; and to the work of Rosenmuller, published at Weimar in 1804, in folio, with engravings of almost all the bones composing the skeleton of the extinct bear, the size of which approached nearly to

o 2

that of a large horse.   M. Goldfuss also in his Taschenbuch has given a general description of all the most important caves that occur in the neighbourhood of Muggendorf; and Leibnitz in his Protogæa, and De Luc in vol. 4 of his Lettres Physiques et Morales, have given considerable details as to the interior of the most important caverns in the Hartz.

M. Rosenmuller says he has never seen the remains of the elephant and rhinoceros in the same cavern with those of bears ; but that he has found the bones of wolves, foxes, horses, mules, oxen, sheep, stags, roebucks, badgers, dogs, and men* ; and that the number of all these is very small in proportion to that of the bears.   The bones of all kinds occur in scattered fragments ; one entire skeleton only of the Ursus spelæus is said to have been found by Bruckmann, in a cave in the Carpathians, and to have been sent to Dresden†.   He adds, that the different state of these bones shows that they were introduced at different periods, and that those of all the animals last enumerated, including man, are in much higher preservation than those of the bears and hyænas.

Thus it appears that the bones which are in most perfect preservation, and belong to existing species, have been introduced during the post-diluvian period; whilst the extinct bears and hyænas are referable to the antediluvian state of the earth.   In corroboration of

* Mr. Esper has found in the cave of Gailenreuth many fragments of sepulchral urns, which from their form were probably made at least 800 years ago: they were of four kinds, and some of them must have been two feet in diameter, others much smaller.

† Being at Dresden myself last summer, I ascertained that no such skeleton exists at present in the royal collection in that city, although there are magnificent specimens in it of heads and single bones of the fossil bears and fossil elephants.

this, I found in 1820, in the collection of the Monastery of Kremsminster, near Steyer, in Upper Austria, skulls and bones of the Ursus
spelæus in consolidated beds of diluvial gravel, forming a pudding-
stone, and dug for building near the monastery.   M. Cuvier mentions the occurrence of teeth, supposed to be those of bears, with the
remains of elephant, rhinoceros, and hyæna, in the diluvium near
Canstadt, on the Necker; and Mr. Pentland has discovered in Italy
the remains of bears mixed with the bones of hyæna, elephant, and
rhinoceros in the diluvium of the Val d'Arno.   Hence it appears that
these bears lived with the elephant and rhinoceros in the period immediately preceding the formation of the diluvium; and the same
thing has been already shown of the extinct hyæna in the gravel of
France, Germany, and England.

M. Rosenmuller states that in all the caverns he has examined, the
bones are disposed nearly after the same manner; sometimes scattered
separately, and sometimes accumulated in beds and heaps of many
feet in thickness; they are found every where, from the entrance to
the deepest and most secret recesses; never in entire skeletons, but
single bones mixed confusedly from all parts of the body, and animals
of all ages.   The skulls are generally in the lowest part of the beds of
bone, having from their form and weight sunk or rolled downwards,
through the longer and lighter bones, during the agitation to which
they have been submitted; the lower jaws are rarely found in contact
with or near to the upper ones, as would follow from the fact last
mentioned*.   He adds, that they are often buried in a brown

---

* At Kirkdale, not one skull, and few, if any, of the larger bones, are found entire;
for these had all been broken up by the hyænas to extract the brains and marrow; and
in their strong and worn out teeth we see the instruments by which they were thus

argillaceous or marly earth, as in the cases of Gailenreuth, Zahnloch, and in the Hartz, and that some of this earth, from an analysis by M. Frischman, was found to contain a large proportion of animal matter.

In the caves of Gailenreuth and Mockas, a large proportion of the bones is invested with stalactite. Even entire beds, and heaps of them many feet thick, are sometimes cemented together by it, so as to form a compact breccia, but they are never found in the substance of the rock itself. At Sharzfeld and in the Carpathians, they are sometimes enveloped with agaric mineral (lac lunæ); they have undergone no alteration of form, but the larger bones are generally separated from their epiphyses. Their usual colour is yellowish white, but brown where they have lain in dark-coloured earth, as at Lichtenstein. At Mockas their degree of decay is by far the greatest; even the enamel of the teeth is far gone, and the bones are perfectly white, having lost all their animal gluten, and acquired the softness and spongy appearance, as well as colour, of calcined bones; still their form is perfect, and substance inflexible, and, when struck, they ring like metallic bodies falling to the ground. These retain simply their phosphate of lime. In other caverns they are usually less decayed, but they sometimes exfoliate and crack on exposure to air, and the teeth, particularly, are apt to split and fall to pieces, as are also those at Kirkdale*.

---

destroyed. The bears, on the other hand, not being exclusively carnivorous, nor having teeth fitted for the cracking large bones, have left untouched the osseous remains of their own species.

* It is mentioned as a curious accident, that of five caves in the calcareous hills, near Muggendorf, that flank the valley of the Weisent-stream, three on the north chain contain not a fragment of any bones, while two on the south side are full of them. This

M. Rosenmuller is decidedly of opinion with M. Cuvier, that the bears' bones are the remains of animals which lived and died through successive generations in the caves in which we find them; nay, even that they were also born in the same caves; in proof of which he has found some bones of a bear, so small, that it must have died immediately after its birth, and other bones of individuals that must have died in early life, like the young hyænas which have been found at Kirkdale: and M. Blumenbach expresses precisely the same opinion in his first Specimen Archæologiæ Telluris, p. 14.

" Utut interim sit, speluncarum istarum ratione et ossium ursinorum in iisdem situ persuasus sum, ea, neque ut quorundam ferebat opinio ab hominibus illic illata, neque quæ aliorum sententia fuit a diluvio illuta esse, sed quod et Cl. De Luc asserit, ipsos istos specus harum ferarum nativos quondam recessus et postmodum sepulchretum fuisse."

The above description of the cave at Gailenreuth, extracted principally from Rosenmuller, and confirmed by my own observations on the spot, may be taken as an example of the general state of the bones in the other caves on the Continent, of which it is superfluous here to say any thing more than to subjoin a list of the most important of them,

may probably be explained by supposing the mouths of the former to have been closed and inaccessible in the antediluvian period, and afterwards laid open by denudation. In the adjacent valley of the Esback, at the castle of Rabenstein, they occur in caves on both sides of a similar valley of denudation; and even admitting them to have been all open and accessible in the period above alluded to, it does not follow that they would all have been equally tenanted by bears, whose gregarious habits would lead them to prefer a frequented den to a solitary one; and this predilection, acting through successive generations, would accumulate the bones of hundreds or thousands in one cave, by the side of another which may have remained all the while almost wholly unoccupied.

and refer to M. Cuvier's Animaux fossiles, for further details taken from the authors by whom these caves have been described.

The caves alluded to are as follows:

1. That of Scharzfeld, in Hanover, in the south border of the Hartz, described by Leibnitz, De Luc, and Bruckmann.

Behrens, in his Hercynia Curiosa, speaks of several more in the neighbourhood of the Hartz; from most of these the bones were collected during a long course of years, and sold for their imaginary medicinal virtues, as the bones of the Licorne, or fossil Unicorn, of which a most absurd drawing is given by Leibnitz in his Protogæa.

2. That of Bauman, in the county of Blakenberg, in Brunswick, on the east border of the Hartz forest, and described by Leibnitz and De Luc.

3. The caves that next attracted attention were those of the Carpathians, and the bones found in them were at first known by the name of dragons' bones, and have been described by Hayne and Bruckmann.

4. But the most richly furnished are the caves of Franconia, described by Esper and Rosenmuller, near the sources of the Mayn, between Nurenburg, Bamberg, and Bayreuth, in the vicinity of Muggendorf, and known by the names of Gailenreuth, Mockas, Zahnloch, Zewig, Rabenstein, Schneiderlock, and Kühloch.

5. A fifth locality occurs at Glücksbrun and Leibenstein, near Meinungen, on the south-west border of the Thuringerwald.

6. And a sixth in Westphalia, at Kluterhoehle and Sundwick, in the country of Mark. An account of these is shortly to be published by Professor Goldfuss and Mr. A. L. Sack, of Bonn.

M. Cuvier in his first edition states, that the bones found in these caverns are identical over an extent of more than 200 leagues; that three-fourths of the whole belong to two species of bear, both extinct; the ursus spelæus and ursus arctoideus*, and two-thirds of the remainder to extinct hyænas†; a very few to a large species of the cat family, being neither a lion, tiger, panther, or leopard, but most resembling the jaguar of South America: with them is found a species of glutton, and a wolf or dog (not distinguishable from a recent species), a fox, and polecat.

It has been said, that in the caves thus occupied, there occur no remains of the elephant, rhinoceros, horse, ox, tapir, or any of the ruminantia or rodentia, and in this respect they differ materially from that of Yorkshire; but such variation is consistent with the different habits of bears and hyænas, arising from the different structure of their teeth and general organization; from which it follows that bears prefer vegetable to animal food; and, when driven to the latter, prefer sucking the blood to eating the flesh, whilst hyænas are beyond all other beasts addicted to eating bones.

From this circumstance it is indeed probable, that in the caves

---

* M. Soemmering, whom I visited at Frankfort last summer, has in his collection the head of a third species of bear, from the cave of Guilenreuth, not distinguishable from the head of an existing species of brown bear from North America, which he had placed by the side of it for comparison. He had the kindness to lend me this specimen, that I might convey it to M. Cuvier at Paris. It was much calcined, and appeared to be of the same age with the bones of the extinct species.

† I am disposed to think that this proportion, as it relates to the hyænas, is too large; for in visiting all the caverns, as well as several collections of bones taken from them, in Germany, I could find very few fragments of teeth or bones of the hyæna, amidst hundreds which belonged to bears.

P

inhabited chiefly by bears, the bones of other animals should be extremely rare; they are not, however, wholly wanting. M. De Luc (Lettres, vol. iv. p. 588) mentions that the remains of rhinoceros have been found in the cave of Scharzfeld, and ascertained by Professor Hollman of Göttingen, to belong to that animal; and M. Soemmering informed me, that he is assured of the fact, having seen the tooth of the animal here alluded to in Hollman's collection at Göttingen: an account of it is given by Hollman in the Comment: Göttingens: M. Esper also mentions, that bones of elephants have been found by M. Frischmann in the cave of Schneiderloch; and speaking of Zahnloch, his words are, "On a trouvé ici des morceaux des dents d'elephant, ce que les fragmens prouvent incontestablement, la forme, la croissance, la structure interne, et en général tous les characteres, mettent hors de doute la réalité de cette production." And Mr. Sack, of Bonn, whom I have before quoted, has found in the cave of Sundwick, within the last two years, the molar teeth and bones of the foot of rhinoceros, and the horns, jaws, and other bones of deer, in the same cave with the remains of the hyæna, glutton, and two extinct species of bear; they have moreover the marks of teeth on them, and their softer parts and condyles have been gnawed off. Hence it follows, that graminivorous animals occur, though rarely, in the caves of Germany; and they may either have been washed in together with the diluvial loam and pebbles, or have been dragged in for prey by the few hyænas that occasionally intruded. That the elephant and rhinoceros lived in the neighbourhood of these caves, in the period immediately preceding the formation of the diluvium, is probable, from the abundant occurrence in it of the bones of both these animals near the caves of Scharzfeld, in the Hartz, and of Altenstein, in Saxe Meinungen, mentioned by Blumenbach. (Archæologia Telluris, Part I. p. 13 and 15.) We shall hereafter see that carnivorous animals have been found recently in vertical fissures, as well as in the caves.

Professor Hollman, in the Commentar: Göttingens: for 1752, T. II. p. 215, has published an account of many baskets full of these large bones, which were discovered 70 years ago, in marl pits (i. e. diluvial loam), at the village of Horden, near Herzberg, and within six miles of the cave of Scharzfeld, and sent to Göttingen; amongst them were portions of five skeletons of rhinoceros; and I have already mentioned another discovery of the bones of elephant, rhinoceros, and hyæna, made in 1808, in the same neighbourhood of Herzberg, between Osterode and Dorst; they were also embedded in diluvial marl, and are described by Professor Blumenbach, in Part II. of his Archæologia Telluris. The facts of the same extinct species of hyæna, being common to the caves and gravel of France, Germany, and England, and of bears occurring in the diluvial gravel of Upper Austria, Wirtemberg, and Italy, prove both these animals to have been the antediluvian contemporaries of the extinct elephant and rhinoceros; there is, therefore, no anachronism in finding the remains of the two latter in dens that were occasionally inhabited by the hyænas and bears.

EVIDENCE OF DILUVIAL ACTION IN THE CAVES AND FISSURES
OF GERMANY.

I come now to consider, what is the evidence of diluvial action afforded by these caverns, and how far it is analogous to that which we find in the caves of our own country; and having made it my business during the last summer to visit Germany, for the purpose of investigating this important question, I shall now proceed to show by a detailed description and drawings of the interior of those among them which are most remarkable for containing bones, that there prevails throughout them all, in comparing them with each other, as well as with those in England, a harmony of circumstances exceeding what my fullest expectations would have anticipated; all tending to establish the important conclusion of their having been once and once only submitted to the action of a deluge, and that this event happened since the period in which they were inhabited by the wild beasts. In every cave I examined, I found a similar deposit of mud or sand, sometimes with and sometimes without an admixture of rolled pebble sand angular fragments of rock, and having its surface more or less abundantly covered over with a single crust of stalagmite; and in those among them, which had been inhabited as dens before the introduction of the mud and pebbles, the latter are always superinduced upon the remains of the wild beasts.

I had, indeed, in my former edition, extracted the same conclusion from descriptions given by De Luc, Esper, Leibnitz, and other writers; parts of which I have subjoined in a note below. But they had, many of them, overlooked the fact of the occurrence of pebbles in the earthy sediment, and the no less important feature of there being but one stalagmitic crust incumbent upon the mud; and they had ventured to offer no reasonable conjecture, as to the time or manner of the introduction of these earthy materials, or their relations to the period in which the caverns were inhabited; these desiderata I am now enabled to supply on my own authority, having conducted my observations with a careful regard to a comparison of the phenomena of these caverns, with those that occur in England *.

The caverns themselves are composed of a succession of vaulted chambers, communicating with each other by long and narrow passages, ascending and descending irregularly through limestone rocks

---

* De Luc has described in the following words, the matrix in which the bones are lodged in the cave, at Scharzfeld: "Le fait est donc simplement, que le sol de ces caverns est d'une terre calcaire, qu'en creusant cette couche molle, on en tire quantité de fragmens d'os, et qu'il s'y trouve aussi des concrétions pierreuses qui renferment des os."—De Luc, Lettres, Vol. IV. p. 590.

Leibnitz, also, in his description of this cavern, has the following passage to the same purpose, "Limo nigricante vel fusco infectum est solum."—Leibnitz, Protogæa, p. 65.

Esper thus describes the state of the floor near the entrance of one of the largest caverns at Gailenreuth. "Dans toute la contrée, le terrain est marneux, mêlé avec du limon, et tire sur le jaune, mais ici on trouve une terre moins limoneuse dans une profondeur considérable."—Esper, p. 9.

Rosenmuller, speaking generally of the same subject, says, "Ces fragmens se trouvent assez souvent déposés dans une terre brunâtre argileuse ou marneuse, comme dans les cavernes près de Gailenreuth au Zahnloch, et dans les cavernes du Hartz." It is also stated, that a sediment of mud was found on the sides and floor of a cave, at Glucksbrun, in the Thuringerwald, near Meiningen, when it was newly opened in cutting a road in 1799; and that in other caverns also there is mud. In all the above quotations, the fact of the mud is clearly stated, but no satisfactory attempt is made to offer any explanation as to its origin. My own observations in the last summer enable me now to speak with more confidence and precision on this subject than I could do on the authority of others.

of various ages and formations. The general state of their interior is nearly as follows. 1. The first thing we see on entering them is an irregular carpet or false floor of stalagmite: this has been much broken, and almost wholly destroyed, in those which have been ransacked for centuries in search of bones; but in the newly discovered caves, and in others, which, containing but few bones, have not been broken up, its extent is great, and sometimes total. 2. Between this crust, and the actual floor of native rock, there is usually a bed of loam or diluvial mud, interspersed with rolled pebbles, angular stones, and bones, and varying in thickness from a few inches to 20 or 30 feet; there is no alternation of the mud or pebbles, with any second or third general crust of stalagmite, nor any thing to indicate that the cause which introduced them has operated more than once. 3. Beneath this mud, we arrive at the native rock, or actual floor of the den, the surface of which is very uneven, and sometimes polished, as if by the trampling of its antediluvian inhabitants.

In those caverns, which appear to have been occupied as dens of wild beasts, before the introduction of the mud, the quantity of bones contained in the uppermost chambers is comparatively small; but, as we descend deeper, we find them more and more abundant, till, at length, in the lower vaultings, or cellarage, they are accumulated in enormous heaps, and the vaults themselves become filled and entirely choked up with a congeries of bones, pebbles, angular stones, and mud, piled confusedly together. In many portions of this mass the earth is loose, and the bones may easily be extracted; in other parts it is consolidated by stalagmitic infiltrations into a hard osseous breccia, resembling that of Gibraltar, and along the shores of the Mediterranean, but not so red. It resembles it also in being full of irregular cells, and of small veins, that are lined internally with a thin pellicle of stalagmite. In this breccia of the under vaultings artificial holes, or small galleries, have been dug to extract the bones; and of these only it is true that the roof and sides, as well as the floor, have bones adhering to them: in the natural chambers there is not a single fragment of bone, except upon or below the floor.

These general observations apply to the caves and fissures near Spa, as well as to those of the Hartz Forest and Franconia, and it will be convenient to begin my more detailed account of them with those cases that are most simple.

I. CAVES NEAR SPA.

In the transition limestone which occurs in the neighbourhood of Spa, at Theux, and Verviers, I found numerous vertical fissures extending upwards to the surface, and often communicating laterally with other fissures and with small caverns. All these fissures were filled entirely, and the caverns partially, with a mass of diluvial mud and rolled pebbles. Amongst the latter were chalk, flints, and many varieties of quartz and slate rocks. The mud in some of the larger fissures abounds with ochreous concretions, formed stalagmitically since its introduction to its present vertical position; and like that in the cave near Wirksworth, illustrates the history of the ochre I have before mentioned, as having been worked in a similar fissure containing antediluvian bones of elephants, &c. at Hutton, in the Mendip Hills. In the caverns near Theux, I found a crust of stalagmite covering the mud; and above this stalagmite some bones of modern animals, e. g. chicken, fox, dog, and sheep: the mud beneath the stalagmite has not been examined, but I could obtain no intelligence of bones having been as yet discovered in it. In this district there are other large caves near Spa and Verviers, which I had not time to visit, but which deserve examination, with a view to the discovery of bones beneath their stalagmitic floor*.

II. CAVES IN WESTPHALIA.

I had no opportunity to visit the caves of Klüterhoëhle and Sundwick, in Westphalia, but I was informed that they occur in the same transition limestone as those of Verviers and Theux; and in the Museum of the University of Bonn, I found magnificent heads

---

* The manner in which these fissures are filled up to the surface of the soil is precisely similar to that of the mud veins, or dykes, which I have already described as occurring in the limestone rock at Chudleigh, and to those dykes which Mr. Strangways (in the 5th vol. of the Geol. Trans. Plate 26) has represented as occurring in vertical fissures of transition limestone on the banks of the Pulcovca, near Petersburgh, and as being filled with diluvial gravel, containing boulders of granite, such as are scattered over the surface of all that country: these dykes have immediate communication with a bed of the same diluvium, also containing granite boulders, and reposing on the surface of the limestone in which the fissures occur.

It is probable that the observation of dykes, or veins, of this kind, so evidently filled by substances poured in from above, suggested to Werner the erroneous idea, that basaltic and metallic veins also had been filled by materials introduced from above in a similar manner.

of the fossil bear, tiger, and glutton, collected from thence. These heads have been recently described, with excellent engravings, by Professor Goldfuss, in the Nova Acta Naturæ Curiosorum, v. 10, and are precisely of the same species with those found in the caverns of England, the Hartz Forest, and Franconia. I have already mentioned that further accounts of the Westphalian caves are in preparation by Professor Goldfuss, and M. A. L. Sack, of Bonn.

III. CAVE OF SCHARZFELD.

The next caverns I examined were those of Scharzfeld, near Herzberg, in Hanover, on the west border of the Hartz Forest, not far from Göttingen. The rock in which these caves occur is magnesian limestone, of the same age with the limestone of Sunderland, in England, and being the first floetz limestone of Werner; occasionally it is very cellular, and abounds in fissures and caverns. The position of the great cave containing the bones is at an elevation of at least 500 feet above the nearest river, and in the centre of one of the many wooded ridges which connect the higher mountains of the Hartz with the plain, and are separated from each other by deep valleys of denudation.

The entrance to this cave is not, as usually happens, in the side of a rocky cliff or precipice, but by an open fissure, placed like a well, in the surface of a plain field, and communicating directly downwards by a steep but practicable descent on ledges of rock into the body of

Q

the cave. (See Plate XIV. A, and explanation.) This fissure may
have been the actual aperture by which the animals came in and out
whilst the cave was inhabited, and by which also the mud and pebbles that occur below were drifted in. Descending by it, we find
ourselves in a long and lofty cavern, terminated at one extremity by
the fissure just mentioned, and proceeding in the opposite direction
to a great distance into the body of the hill. It has several lateral
communications and connecting passages, for a detailed description
of which I refer my readers to De Luc's Letters, vol. 4. The point to
which I wish, at present, to direct their attention, is the state of the
floor and position of the bones.

The floor appears to have been at one time covered in many
places with a crust of stalagmite, of which there are a few traces still
remaining, but the greater part has been destroyed by visitors continually digging into the subjacent mud in search of bones. In my
drawing (see Plate XIV. D) this crust is restored to the state in which
it probably existed when the first diggings began to be made in it;
at present we see little more than a bed of mud and pebbles, and
broken fragments of stalagmite, covering the bottom of the cave, and
interspersed with teeth of bears and other animals, and fragments of
bones. In some parts of the floor, holes like that at E have been
dug through this bed of mud to the limestone rock below, for the
purpose of collecting bones; in other parts the natural rock projects
above the surface of the mud, and is without any stalagmite. The
general appearance of the whole corresponds with the description
given of it by Gottschalk in his Guide for the Hartz: "the bottom is

almost every where covered with a fine loose earth, full of broken bones."

Along the edges of this floor are seen a number of smaller caverns, which pass off from the main chamber to various distances in the body of the rock; the bottom of these is filled also with the same materials that cover the floor of the great chamber.   Beneath the latter are also numerous undervaultings, and small branching catacombs of irregular shape; (see H, I;) some terminating in a cul de sac, others communicating by a lateral aperture with some adjacent cavity, which again has further communications either with the main chamber above, or with other smaller ones below, so that the rock is intersected and undermined like an irregular mass of honeycomb.   These undervaultings have for the most part been entirely filled up, as at I, with a mass of brown earth, or diluvial loam, through which, as their matrix, are disseminated enormous quantities of broken bones, teeth, angular fragments and pebbles of limestone. In the cavities thus choked up there is no room for any stalagmitic crust, as there is no expanded surface over which it could be spread. The mass which fills them, however, is in some parts firmly cemented together by stalagmitic infiltrations of calc sinter; more frequently the mud is semi-indurated, spungy, and cellular, and may be readily cut with a knife; in other parts it is quite soft and loose, and the bones and pebbles are simply imbedded in it.

These lower vaults have in no case that I could find been laid open to their full extent, but are still choked up below with matter of the nature above described, and would no doubt richly repay the labour of any person who has leisure to explore them.   The excavations that have been made in them have produced small artificial caverns, the sides and roof of which are crowded with, and sometimes in great part made up of bones; (see Plate XIV. H;) but not a particle of bone occurs adhering to the roof or sides of the great chamber (B) above the level of its floor *.

There are several artificial excavations similar to that represented at H, in one of which I found the head of a large bear, from which the upper part of the skull had been broken off, and a pebble lay imbedded in mud in the cavity of the skull. Close to it was an under jaw, (possibly of the same animal,) and both were surrounded with mud, pebbles, and bones, the latter exceeding in proportion both the former: they were so firmly packed together, that with a small hammer and chisel, I could advance but slowly in extracting them; but, with proper instruments, cart-loads of bones might easily be obtained. Whilst working on this mass, I could not help imagining that I was in the cavernous fissure at Plymouth, so precisely analogous were all the circumstances before me to those I had there witnessed, as to the manner and matrix in which the bones were packed together, and the difficulty of unpacking without destroying them. Their state of preservation also is nearly the same, being less perfect than is usual in the other caves of Germany and England. Instead of the ordinary

---

* I dwell more on this circumstance than I should have otherwise thought necessary, because preceding writers on this subject have overlooked the important distinction I am now drawing, and have stated the occurrence of bones in the roof and sides in such general terms, that persons who have not themselves carefully inspected the caverns with a view to this particular point, may conclude from their descriptions that the bones are found indiscriminately adhering to the sides and roof of the upper chambers, as well as of the lower vaults, which is not the fact.

white or yellow, they are of a dingy brown colour approaching to black; and whilst wet may be readily crushed to a dark earth-like powder, which I presume to be the black earth said by many writers to abound in this cavern, but of which I could neither here, nor in any other cave, excepting that of Kühloch, near Muggendorf, discover a single particle, though I looked for it very carefully in every direction. The circumstances of this exception are very peculiar, and will presently be described. The looseness of the earth in which the bones are for the most part embedded, both at Scharzfeld and Plymouth, and the comparatively small quantity of stalagmite that accompanies them, may explain the cause of their greater state of decay than is usual where the mud is more argillaceous, or the incrustation of each individual bone with stalagmite more complete.

IV. CAVE OF BAUMANS HÖHLE.

This celebrated and much frequented cave, or suite of caverns, has already been described by Leibnitz in his Protegæa. It derived its name from an unfortunate miner, who, in the year 1670, ventured alone to explore its recesses in search of ore; and after having wandered three days and nights in total solitude and darkness, at length found his way out in a state of such complete exhaustion, that he died almost immediately. It lies in a bed of transition limestone at the village of Rubeland, about two miles below the town of Elbingrode, on the north-east border of the Hartz, and in the country of Blankenburg; its relative position to the nearest river is low when compared with Scharzfeld, being at an elevation of about 100 feet above the bed of the Bode; this, however, is sufficient to assure us, that it is quite impossible to attribute the pebbles and mud within this cave to any floods of that river, which could not rise ten feet without destroying the adjacent village of Rubeland. Above this village is seen the present entrance of the cave, in a nearly per- pendicular cliff, which forms the left side of a deep gorge, through which the river runs, and which is similar to that through which the Avon passes at Clifton. (See Plate XV. L.) The breadth of this gorge varies from 100 to 300 feet, its depth is about 150; the rocks on both sides of it are nearly precipitous. In the drawing the scale is falsified as to breadth, for the sake of getting room *.

The present entrance (see Plate XV.) is by the aperture A into a low flat cavern, 15 feet broad, and five feet high; the outer extremity of which is in the truncated face of the cliff, whilst within it descends rapidly to the broad and lofty chamber B. The form of this chamber is irregularly oblong, varying from 30 to 50 feet in diameter, and from 10 to 20 in height, and affording some of the most grand and picturesque features of cave scenery. The floor of this cave resembles in all its circumstances that of the great cave at Scharzfeld, with the addition of several large masses of rock (see O, O) that have fallen

---

* This gorge is simply a valley of denudation, produced (like hundreds which I could mention in compact limestone countries) by the force of the diluvian waters. The dotted lines N (Plate XVI.) represent the manner in which the two sides were probably connected before the excavation of the gorge, and M the probable continuity of the pre- sent mouth of the cave A to the then existing surface. As I shall hereafter consider the evidence geology affords of the formation of such valleys by the diluvian waters, I at pre- sent beg those of my readers to whom this subject is new to allow the fact to be as- sumed in the case immediately before us.

from the roof, and stand high above the surface of the mud and broken stalagmite, which is represented as restored at c.   From the great cave B, we descend by the passage D to the hollow vault E, the lower half of which contains beneath a thick crust of stalagmite, an accumulation of several feet of mud or sand mixed with bones, and extremely large pebbles of transition limestone; the mud and pebbles have been separated from each other, and drifted to different parts of this vault.   The bones which lie in the mud and sand are not much broken, and about 30 years ago some very entire ones were extracted from it, and sent to the Museum at Brunswick; but those which occur among the pebbles are more than usually fractured, and some of them stamped or pounded, as if in a mortar, into hundreds of small splinters, which adhere by stalagmite to the surface of some of the largest pebbles: none of them, however, have lost their angles, or are in any way rounded; but they are simply broken or crushed when in juxta-position to the heavy pebbles, which are more abundant and larger here than in any other part of this, or indeed of any cavern I have yet visited.

It follows from these facts, that the pebbles could not have received their actual state of perfect roundness by violent agitation of water within the cave itself, as in this case the bones also would either have been reduced to pebbles, or totally destroyed; they were probably, therefore, rounded before they reached their present place, being derived from the limestone rocks of the adjacent country, and drifted in by the mouth A and its prolongative M, whilst the valley L was in the act of being excavated.   Thus introduced, they might have passed downwards across the cavern B to the vault E, where the rock F would form an impediment to their further progress; many bones that lay in the great cave B would pass forward with these pebbles into E, and others may have already been laid there; in either case, a rapid movement of the large pebbles must have been necessary to crush to pieces the large and strong bones, whose sharp and angular splinters now adhere to them. Their state again is totally different from that of the splinters in the den at Kirkdale, which latter are as obviously the effect of fracture by the hyænas' teeth, as the former are of a violently crushing blow, imparted by a heavy mass of stone. These splinters, as well as the less injured bones that lie among the pebbles, are completely cased with stalagmite, by which they adhere to, and form with the pebbles a strong breccia; their preservation from decay is complete, and their colour that of natural bone, or light cream colour.

Within the vault E, the rock F rises almost suddenly about 20 feet, and must be crossed by ladders; on the further side of it, we descend again a considerable distance by a lofty and rugged aperture to the lower cavern H, from the roof and sides of which there ascend other passages resembling K, which I did not explore, but of which, and indeed of the whole cavern, a ground plan is given by Leibnitz, in his Protogæa, Plate I. By some of these passages, the pebbles may have come in which we find below, in the cavern H. This cavern has, from its position in the inmost recesses, and its difficulty of access, been not much disturbed, and has several off-shoots, the contents of which are still glazed over with a crust of virgin stalagmite: in others, the stalagmite has been broken through as at I; and artificial vaults, like those at Scharzfeld, have been dug some feet into the subjacent mass of mud, which is also loaded with teeth, bones, and pebbles, but not with such large pebbles, or in such unusual quantity as in the vault E.   The rock and sides of the artificial cave I have bones adhering to them, or rather are in part composed of bones; but in none of the natural chambers do we find bones adhering to the side and roof above the surface of the mud and stalagmite.

All these circumstances are corroborative of the hypotheses I am endeavouring to establish.   First, That the agent, by which the mud and pebbles were introduced, was the same diluvial waters, which extirpated the animals that had antecedently inhabited the cave. Secondly, That this diluvial detritus was not introduced at different intervals by the action of rivers, or land-floods, but was by one single operation superadded to the bones already existing in the dens. Thirdly, That the period of its introduction is that from which we must begin to date the formation of the superficial crust of stalagmite, by which these diluvian and antediluvian records have been sealed up, and maintained in such high preservation to the present hour.

CAVE OF BIELS HÖHLE.

This cavern is said to have derived its name from a heathen temple, that formerly stood on the edge of the cliff immediately above it, and of which some traces still exist: its position is at the distance of but a few hundred yards from the Baumans Höhle, last described; and at nearly the same elevation in the cliffs on the opposite side of the gorge, E, and on the right bank of the Bode river. (See Plate XVI.) No bones have been as yet discovered in it, nor does it contain any such lofty and broad chambers, as those at Scharzfeld, and Baumans Höhle : it is composed of a succession of cavernous vaults, ascending and descending irregularly in the transition limestone, and communicating with, and intersected by, other similar tubes, which traverse the body of the rock in various directions (see Plate XVI. c. c. c.) It is remarkable for the beauty of the stalactite that hangs from its roof, and the quantity of stalagmite that forms an universal cover of considerable thickness over its floor; but it has not as yet been discovered to contain bones. In passing through it, we are obliged to ascend by ladders a succession of rocky projections or pinnacles (see Plate XVI. D. D. D.), between which we descend into as many intervening hollows or basins of unequal size and height, B. B. B. ; and having entered by the small hole A in the cliff, overhanging the River Bode, we come out by an artificial hole in the same cliff, at a small distance from the former. Other cavities of the same kind rise probably to the surface of the land by the tubes c. c. c., and are choked up by diluvium. The dotted lines F. G. represent the restoration of the rock and of the tube A to the state in which they probably existed before the excavation of the gorge E.

This cave appears to be one of those which has never been inhabited as a den by wild beasts, as had it been so, it is probable some traces of bones would have been found in the excavations which have been dug in its floor, for the purpose of making an easy path for visitors, that crowd to see the beauty of its stalactite. To myself, these excavations afforded matter of much higher interest, as they enabled me to identify beneath the crusts of stalagmite H. H. the same bed of diluvial mud I. I., which I had already seen at Baumans Höhle, and Scharzfeld, and Theux, and in the caves of England, and alternating also as it does at Plymouth, with thin beds of blue clay, and coarse loose sand, mixed abundantly with small fragments of greywacke slate and clay slate. I was told, though I did not see them, that it sometimes contains pebbles. It is remarkable, that this diluvium is accumulated on the top of the pinnacles D. D. D., in nearly as great quantity as in the intermediate basins B. B. B; and here again, we have another analogy to the caves at Plymouth, viz. that wherever there was a ledge, or shelf, or basin, however minute, whereon there was space enough for the smallest deposit to take place, from a mass of water loaded with mud, and sand, there these materials have found a lodgement, and have ever since remained undisturbed, under a gradually accumulating crust of stalagmite. This single post-diluvian crust H. H. is the only one that appears both in the basins and on the pinnacles: it is spread on the upper surface only of the diluvial sediment I. I.; and in no case has it been found to alternate with the mud, or sand, or clay, of which this sediment is composed. I was unable to ascertain, whether or not there be (as at Kirkdale), a subjacent crust of stalagmite, accumulated on the native rock beneath, before the introduction of the mud. On the surface of the upper stalagmite there is a quantity of mould, which has been brought in by the present guide from the adjacent field, for the purpose of making a path in the interior of the cave.

R 2

CAVES IN FRANCONIA.

Having thus far ascertained, by careful examination of the most important caverns in the Hartz, that there exists no discrepancy, or rather a complete analogy in all their phenomena to those of the caves in England, allowing as much as is due to the different habits of the animals whose remains we find in them respectively, my next attention was directed to the no less celebrated assemblage of caverns in Franconia, situated in the district round Muggendorf, in the Upper Palatinate, about 30 miles N. E. of Nurenberg, in the direction of Bareuth, and nearly in a central point between the three towns of Nurenberg, Bareuth, and Bamberg, (see Map, at Plate XIX.)   The position of this district is at one of the great water heads of central Europe, from which the streams descend on one side southwards, by the Naab and Danube, to the Black Sea, and on the other, by the Mayn and Rhine, to the German Ocean. It is composed of an assemblage of limestone beds, corresponding with the lias, oolite, green sand, and chalk formations of England, and with the Jura and Alpine limestones of the Continent; and forms the important link by which the Jura chain, and its continuations in the Rough or Swabian Alps, are connected with the same formations in the north of Germany, where they are designated by the Wernerian appellation of first and second floetz limestone, and Muschel kalk. The limestone of the immediate vicinity of Muggendorf has been, from its cavernous nature, locally designated by the name of Höhlen kalk (Hole limestone); and it seems more nearly allied to the calcareous portions of our green sand formation than to any other strata of the English series; but differs from them in aspect, in consequence of its containing a large admixture of magnesia, and very few organic remains. This district is at present for the most part cultivated, and without any other beasts of prey than foxes and a few wolves; but there appears to have been a time when its savage population was prodigiously great, and the bones of thousands of gigantic bears, and other wild beasts, which once swarmed in the caverns, with which its hills are perforated, still remain to attest their antediluvian dominion over it.

Though at a high elevation, this district cannot be said to be mountainous; its valleys indeed are flanked by precipitous crags, which, when seen from below, have all the ruggedness and picturesque form of Alpine scenery; but they are narrow, and not deep, rarely exceeding 300 feet, and are simply valleys of denudation, excavated by the diluvian waters on the surface of an elevated calcareous plain, which was the scene on which the bears and hyænas that are entombed in the recesses of its caverns ranged unmolested by the approach of man.

The entrance to the caverns at that time would have been by apertures on the surface of this plain, like that described as now existing at Scharzfeld, and of which some are found still open in the more elevated parts of this neighbourhood; and it seems to have been by these original apertures and fissures rising to the level surface of the main table-land, rather than by the present mouths in the vertical face of the cliffs, that the pebbles, and probably great part of the mud that occurs within the caves, found their admission: these mouths must have been closed, and have formed part of the lower regions of the caverns, before the matter which once filled the valleys had been swept away. When viewed on a correct map (see Plate XIX.), or looked at from the summit of a distant hill, these valleys appear but as open gutters on the surface of a meadow, falling into a main gutter, by which the whole of their waters are carried off, in the same manner as the streams of the Esbach, the Weissent, and others pass off by the main valley of Muggendorf. The manner in which the present mouths of the caverns appear in the cliffs that flank these valleys will at once be understood by referring to Plate XVIII., c, e, i, where three of them are represented in the cliffs that flank the valley of the Esbach, near the castle of Rabenstein, and to the map at Plate XIX., in which the place of each cavern is correctly marked, and a view given of the mouth of the cave of Gailenreuth.

From these preliminary observations on the district, I proceed to describe in detail a few of the most important caverns which it contains. I shall select five, and treat of them in the following order; first, Forster's Höhle; second, Rabenstein; third, Zahnloch, fourth, Gailenreuth; fifth, Kühloch.

1.—FORSTER'S HÖHLE.

This cave is situated near the village of Weischenfeld, in the steep and rocky slope that forms the right flank of the valley of the Zeubach. It was not till within these few years that it attracted

any attention, or was indeed accessible without great difficulty, its
only opening having been a hole in the roof, through which it was
necessary to be let down by ropes or a long ladder.   More recently
a lateral adit had been driven into it by an innkeeper named Forster,
nearly on the level of the floor, and forms the present entrance.
This cave is one of the most remarkable I have ever seen, for the
beauty of its roof, and perfection of its stalagmite; but contains no
other bones than a few of dogs and modern animals, which have
recently been placed in it for the sake of ornament.   Its height
varies from 10 to 30 feet; its breadth where widest being about 30.
Its floor is no where quite level, and the subjacent rock is buried
under a deep bed of diluvial mud, over which is superinduced a more
than usually thick covering of stalagmite.   The form of this stalagmite is particularly striking in the side vaults that descend in various
directions into the main chamber, being inclined at an angle of about
45°, and having the lower half of their area filled also with mud; on
the surface of which entire bridges of thick stalagmite are formed
across from side to side, presenting the varied features and irregular
undulations of large and beautiful cascades, suddenly congealed into
a mass of transparent alabaster: large waving streams of this orna-
mental substance are seen descending into the main chamber from
all the lateral avenues by which it is encircled, and contributing, as
it were, to swell the stalagmitic lake that occupies its centre.   The
roof also of the main chamber, as well as of its side aisles, is in all
parts broken into, and clustered over with irregularly grotesque
forms of exquisite beauty, rivalling the richest combinations of the

most complicated gothic fret-work, and far surpassing them in the wild and irregular varieties in which its masses descend, like inverted pinnacles, to meet the icy lake of stalagmite that covers the floor. The quantity of stalactite dependant from this roof is comparatively small; and though extremely beautiful, it forms a subordinate feature only in this most magnificent vault; the peculiar beauty of which consists in the deep carious cavities into which its entire surface has been corroded, and the endless succession of sharp points and ridges, and irregularly projecting partitions, that stand in high relief between these cavities, and descend many feet, suspended almost by nothing, into the body of the cavern.   With respect to the muddy sediment that occurs below the stalagmitic floor, it remains only to observe, that its thickness has not been ascertained, though several holes have been dug in it to the depth of three or four feet: it contains numerous angular fragments of limestone, but I could find no pebbles.   I have already stated that there are no bones: it probably was inaccessible, or at least not tenanted at the period when the bears were in occupation of this country.   By the holes dug into this mud it appears that there is no trace of any other crust, or even of a film of stalagmite alternating with it, so as to lead us to infer that it was not deposited at one and the same period, and by the same inundation that introduced a similar sediment into all the other caves of which I have been speaking.   Whether there be any antediluvian stalagmite between it and the subjacent rock, I could not ascertain.   There is not a particle of mud, or even of dust, above the surface of the upper crust of stalagmite, or interposed between its component laminæ;

which would have happened, had the mud been either introduced by successive land-floods, or been composed of dust falling continually to the floor from the decomposition of the roof.

2.—CAVE OF RABENSTEIN.

The cave to which I have assigned this title stands near the upper part of the vertical cliff on which is built the castle of Rabenstein, and immediately under the adjacent chapel of Klaustein, by which name also it is sometimes designated. Its position and present mouth of entrance are represented in Plate XVIII. c. It is less remarkable for bones than the three I am next about to mention ; and to this circumstance it owes the preservation of a large proportion of the thick stalagmitic crust that is spread over the mud, which here also is found to cover the floor, and to be mixed with pebbles and angular fragments of limestone, and a few bones and teeth of bears and other animals. As similar remains are so much more abundant, and more easily accessible in other adjacent caverns, these few have not afforded sufficient inducement to the searchers for bones to rip up the stalagmite which covers the diluvial sediment, in which they are here rather sparingly dispersed. The depth of this sediment varies from two to six feet, and upwards, and it differs in nothing essential from that already and hereafter to be described in other caverns. In the few places where the actual floor was visible beneath it, there was no subjacent crust of stalagmite. On the surface of the upper crust I found scattered loosely the recent bones of sheep, dogs, foxes, and some smaller animals. These were not invested with stalagmite, and were in the same state with the recent bones discovered in the modern fissure at Duncombe Park.

3.—CAVE OF ZAHNLOCH.

The next cave I propose to speak of is that of Zahnloch (or the Hole of Teeth), so called from the abundance of fossil teeth that have been extracted from it, and being situated about two miles on the south-east of Rabenstein. Its position is not like that of the other caverns I have to describe in this neighbourhood, in a cliff that flanks some one or other of the valleys, but near the summit of a hill called Höhen Mirschfeld, which rises above the main table-land, and forms one of the most elevated points of this district, being about 600 feet above the valley of Muggendorf. (See Plate XIX.) The entrance of this cave is a low oven-shaped aperture, in a very small rock, that projects through the grass in the north slope of a green and naked hill. It is visible at a considerable distance, and must have attracted notice from the earliest times: it is about ten feet broad, and four feet high, and leads immediately into an extensive crypt-like chamber, about 60 feet in length, and varying from 20 to 40 feet in breadth, but so low that there are few parts in which it is possible to stand upright. From the edge of this larger chamber there branch off several smaller under vaultings, and on one side of it is a cavern the height of which is more considerable than that of the main chamber, and in the middle of which stands a large insulated block of stone, rising about six feet above the present floor, and remarkable for having its surface smoothed over (as if it had been polished), in a way which the natural rock never presents. There is much dripping of water, but very little stalactite hanging from the roof and sides of any of these chambers, and probably the floor also has never been invested with much stalagmite.   At present it is strewed over, to the depth of several feet, with a mass of brown loam, mixed with numerous pebbles and angular fragments of lime-stone, with teeth also, and fragments of bones of bears, and other extinct animals, and with recent bones of hares, foxes, dogs, and sheep: I found also a fragment of a rude sepulchral urn, but could discover no traces of black earth.   The average depth of this loam must be four or five feet: the entire bulk of it has been again and again dug over in search of teeth and bones, of which it still contains considerable quantities.   Even the smaller under vaultings have been ransacked to their extremities; so that there is no possibility of seeing in its natural state any part of this mixed mass, which now covers the floor.   Still all its phenomena, allowing for the dis-turbances that have taken place by digging, are consistent in every respect with those of the other adjacent caverns: the introduction of the mud and pebbles may be referred, as usual, to diluvial agency; some of the angular fragments may have been washed in, others have fallen from the roof; the teeth and bones of the extinct animals may be referred to the wild beasts that inhabited the cave before the in-troduction of the mud, and the sepulchral urn to the use that has since been made of it for the reception of human remains; whilst the bones of hares, foxes, dogs, and sheep, are derived from animals that

s 2

may at any time have entered spontaneously, or been dragged into it, the mouth being so large, and having no appearance of ever having been shut up.

Its most remarkable phenomenon is the polished surface of the great insulated block of stone, which stands like a sarcophagus in the middle of the side chamber. Not being aware of it at the time I was there, I did not observe this circumstance myself, but Rosenmuller states, and Goldfuss repeats it, that the surface of this block is all over smooth, as if polished (glatt wie polirt) in a manner which must be attributed to some cause external to the rock itself, and which its place and circumstances seem to induce us to refer to no other than friction from the skin and paws of the antediluvian bears *. (Goldfuss Taschenbuch, p. 120.)

* It is only necessary to examine the habits of modern bears as delineated after nature in Ridinger's excellent engravings of wild animals, or to have seen the agility and apparent delight with which the bears in the Jardin du Roi at Paris climb up a tree placed for this purpose in their open den, to feel assured that if the habits of antediluvian bears were at all like those of existing species, such a pedestal as that we are speaking of would have been subjected to continual friction from the ascending and descending of these animals, whilst they inhabited the den : see Ridinger's Plates, No. 31, where he represents the interior of a den of bears, in which most of them are climbing up rocks, and one is in the very act of mounting a pedestal similar to that in Zahnloch. The numbers also in which he represents them as herded together (seven or eight in the same cave), show them to be gregarious, and illustrate the otherwise almost inexplicable fact of the remains of so many hundred bears being assembled together in four or five such caves as those of Scharzfeld, Baumans Höhle, Zahnloch, Gailenreuth, and Kühloch.

CAVE OF GAILENREUTH.

I have already mentioned this cave, as the most remarkable in the country we are now speaking of, for the quantity and high state of perfection of the bones that have been extracted from it, and for the descriptions that have been given of them by Esper, Rosenmuller, and other writers. I had visited it in 1816 with less attention to its minute details, and had then overlooked the circumstance of the bed of mud and pebbles, which I now find to be mixed with the bones, and placed between the stalagmitic crust and native rock, which forms the actual floor of the cavern; in other respects the drawing I have given in my first edition in the Phil. Trans. 1822, differs not materially from that which I have now substituted for it at Plate XVII.

Its mouth is situated in a perpendicular rock, in the highest part of the cliffs which form the left side of the valley of the Weissent River, at an elevation of more than 300 feet above its bed. (See Plate XIX. in the corner of which is a view of it, copied from a print in Esper.) This valley being, as I before stated, simply a valley of denudation, the present entrance could not have been the original one, as it existed before the excavation of the valley; we now enter by an aperture about seven feet high and twelve feet broad in the cliff just mentioned, and close to it observe an open fissure, rising from the cave toward the table land immediately above; this fissure is also represented in the view of the mouth of the cave I have just referred to, and by it, or by other similar fissures, the mud and pebbles we shall

find within were possibly introduced. The form and connexions of
the cave will be best understood by referring to the drawing at
Plate XVII. It consists principally of two large chambers, B and F,
varying in breadth from ten to thirty feet, and in height from three to
twenty feet: the roof is in most parts abundantly hung with stalactite;
and in the first chamber, B, the floor is nearly covered with stalagmite,
C, piled in irregular mamillated heaps, one of which in the centre is
accumulated into a large pillar uniting the roof to the floor. From B
we descend by ladders to a second chamber, F, the floor of which also
appears to have been once overspread with a similar stalagmitic crust:
this, however, has been nearly destroyed by holes dug through it, in
search of the prodigious quantities of bones that lie beneath. The cave,
F, is connected by a low and narrow passage, M, with a smaller cavern,
N, at the bottom of which is a nearly circular hole, K, descending like
a well about twenty-five feet, and from three to four feet in diameter,
into which you let yourself down, as in climbing a chimney, by sup-
porting the hands, feet, and back against the opposite sides. In
descending this hole, we find its circumference to be for the most part
composed of a breccia of bones, pebbles, and loam, cemented by stalagmite: on one side of it is a lateral cavity, L, which is entirely artificial,
and is the spot from which the most perfect skulls and bones have
been extracted in the greatest abundance; the lowest cavity, K K, is
also entirely surrounded with the breccia above described; how much
deeper, or how widely it may extend, has not yet been ascertained.
The roof and the sides of the artificial cavities, K and L, having been
dug in the breccia, are crowded with teeth and bones; but these latter
do not occur in the roof or sides of any of the upper or natural

chambers above the level of the stalagmitic crust that covers their floor; this observation, as I have before mentioned, applies equally to all the other caverns I have been describing, and is important on account of the erroneous statements and opinions which exist on this subject. The floor of the first chamber B has been stated to be at this time almost entirely covered with a crust of stalagmite; on the surface of this crust is a quantity of blackish mould, mixed with ashes and charcoal: the latter being derived from fires that are frequently made to illuminate the cave; the former is vegetable mould, which has been brought in from the adjacent land, possibly for the purpose of making a path over the slippery stalagmite, as has been lately done at Biel's Hole in the Hartz. Through this crust of stalagmite some large holes have been dug resembling that at E, and in these we see a bed of brown diluvial loam and pebbles, mixed with angular fragments of rock, and with teeth and bones; but the latter being less abundant here than in the deeper chambers of the cave, the floor has from this circumstance been for the most part left entire. I could not ascertain the depth of this diluvium; where I saw it, it was three or four feet, but the rock below was still invisible. The bones of bears, that lie loosely scattered over the surface of the stalagmite, and even on the outside of the cave's mouth, are rejected fragments that have been dug out from beneath it, or from the lower cavities; and they are mixed with the recent bones of dogs, sheep, foxes, &c. that have entered in modern times by the open mouth, A.

In the second chamber, C, the diluvium is of the same description as in B, but more abundantly loaded with bones; for this reason it has been more disturbed, and the crust remains entire only in a few places. Its depth appears to be irregular, and in parts extremely great. At ɪɪ a side chamber descends rapidly into the body of the rock, and contains cart-loads of teeth, bones, and pebbles, dispersed through a loose mass of brown diluvial loam, but not united by stalagmite, as in the adjacent cavities, ᴋ and ʟ. In these latter they are firmly cemented together into a compact breccia, and accumulated in a heap of at least 25 feet in depth, from the top, ᴋ, to the bottom, ᴋ ᴋ. The distribution of the component materials of this breccia is irregular: in some parts the earthy matter is wholly wanting, and we have simply a congeries of agglutinated bones; in others, the pebbles abound; in a third place, one half of the whole mass is loam, and the remainder teeth and bones: at ᴋ the top, and ᴋ ᴋ the bottom of the well, pebbles and bones occur mixed together in the same proportion as in the middle regions of it. The state of preservation of the bones, when incrusted in stalagmite, is quite perfect, and the colour yellowish white; those extracted from the loose earth are brown or blackish, but I could no where find any thing like the black animal earth described by preceding observers, as occurring in this and the other caves of which I have already spoken.

All these phenomena are in perfect harmony with those of the other caverns in Germany and England; the upper parts of the existing cave, and probably others, which have been cut away by denudation, seem to have been the lodging places of wild beasts, that lived and died in them in the period preceding the introduction of the mud and pebbles. The diluvial waters rushing, as they could not fail to do, into these caverns, would introduce pebbles and mud, and would also drift downwards to their lowest recesses the bones that lay perhaps more equally distributed than they are at present. Since this event, the accumulation of stalagmite on the surface of the mud, and in the interstices of the hollow masses of bones and pebbles, is the only geological change that appears to have taken place: the limestone rock of the actual floor rarely projects so as to be visible beneath the false floor of diluvial matter and bones with which it is overspread; in one place where it does so, in the low passage m, it is smooth and highly polished, like the pedestal in Zahnloch; but whether from the paws of bears, or the hands and knees of postdiluvian visitors, or the united action of both, I will not venture to determine. I could not ascertain whether there was any stalagmitic crust below the mud, as in the cave of Kirkdale.

CAVE OF KÜHLOCH.

It now remains only to speak of the cave of Kühloch, which is more remarkable than all the rest, as being the only one I have ever seen, excepting that of Kirkdale, in which the animal remains have escaped disturbance by diluvial action; and the only one also in which I could find the black animal earth, said by other writers to occur so generally, and for which many of them appear to have mistaken the diluvial sediment in which the bones are so universally imbedded.   The only thing at all like it, that I could find in any of the other caverns, were fragments of highly decayed bone, which occurred in the loose part of the diluvial sediment in the caves of Scharzfeld and Gailenreuth; ·but in the cave of Kühloch it is far otherwise. It is literally true that in this single cavern (the size and proportions of

T

which are nearly equal to those of the interior of a large church) there are hundreds of cart-loads of black animal dust entirely covering the whole floor, to a depth which must average at least six feet, and which, if we multiply this depth by the length and breadth of the cavern, will be found to exceed 5000 cubic feet. The whole of this mass has been again and again dug over in search of teeth and bones, which it still contains abundantly, though in broken fragments. The state of these is very different from that of the bones we find in any of the other caverns, being of a black, or more properly speaking, dark umber colour throughout, and many of them readily crumbling under the finger into a soft dark powder resembling mummy powder, and being of the same nature with the black earth in which they are embedded. The quantity of animal matter accumulated on this floor is the most surprising, and the only thing of the kind I ever witnessed; and many hundred, I may say thousand, individuals must have contributed their remains to make up this appalling mass of the dust of death. It seems in great part to be derived from comminuted and pulverised bone; for the fleshy parts of animal bodies produce by their decomposition so small a quantity of permanent earthy residuum, that we must seek for the origin of this mass principally in decayed bones. The cave is so dry, that the black earth lies in the state of loose powder, and rises in dust under the feet : it also retains so large a proportion of its original animal matter, that it is occasionally used by the peasants as an enriching manure for the adjacent meadows *.

* I have stated, that the total quantity of animal matter that lies within this cavern cannot be computed at less than 5000 cubic feet; now allowing two cubic feet of dust

The exterior of this cavern presents a lofty arch E, in a nearly perpendicular cliff, which forms the left flank of the gorge of the Es-bach, opposite the Castle of Rabenstein. (See Plate XVIII. E.) The depth of the valley below it is less than 30 feet, whilst above it the hill rises rapidly, and sometimes precipitously, to 150 or 200 feet. This narrow valley or gorge is simply a valley of denudation, by which the waters of the Esbach D fall into those of the Weissent. The breadth of the entrance arch is about 30 feet, its height 20 feet. As we advance inwards the cave increases in height and breadth, and near its inner extremity divides into two large and lofty chambers, both of which terminate in a close round end, or cul de sac, at the distance of about 100 feet from the entrance. It is intersected by no fissures, and has no lateral communications connecting it with any other caverns, except one small hole close to its mouth, and which opens also to the valley. These circumstances are important, as they will assist to explain the peculiarly undisturbed state in which the interior of this cavern has remained, amid the diluvial changes that have affected so many others. The inclination of the floor, for about 30 feet nearest the mouth, (see Plate XVIII. E,) is very considerable, and but little earth is lodged upon it ; but further in, the interior of the cavern G is entirely covered with a mass of dark brown or blackish earth H, through which are disseminated, in great abundance, the bones and teeth of bears and other animals, and a few small angular fragments of limestone, which have probably fallen from the roof, but

and bones for each individual animal, we shall have in this single vault the remains of at least 2500 bears, a number which may have been supplied in the space of 1000 years, by a mortality at the rate of two and a half per annum.

I could find no rolled pebbles. The upper portion of this earth seems to be mixed up with a quantity of calcareous loam, which, before it had been disturbed by digging, probably formed a bed of diluvial sediment over the animal remains; but, as we sink deeper, the earth gets blacker and more free from loam, and seems wholly composed of decayed animal matter. There is no appearance of either stalactite or stalagmite having ever existed within this cavern.

In some of the particulars here enumerated, there is an apparent inconsistency with the phenomena of other caverns, but the differences are such as arise from the peculiar position and circumstances of the cave at Kühloch: the absence of pebbles, and the presence of such an enormous mass of animal dust, are the anomalies I allude to; and both these circumstances indicate a less powerful action of diluvial waters within this cave than in any other, excepting Kirkdale. To these waters, however, we must still refer the introduction of the brown loam, and the formation or laying open of the present mouth of the cavern: from its low position so near the bottom of the valley, this mouth could not have been exposed in its present state, and indeed must have been entirely covered under solid rock, till all the materials that lay above it had been swept away, and the valley cut down nearly to its present base; and as the cave ends inwardly in a cul de sac, and there is no vertical fissure, or any other mode of access to it, but by the present mouth, if we can find therein any circumstances that would prevent the admission of pebbles from without, or the removal of the animal remains from within, the cause of the anomaly we are considering will be explained. By referring to Plate XVIII. it will be seen that the throat of the cave F, by which we ascend from the mouth E to the interior G, is highly inclined upwards, so that neither would any pebbles that were drifting on with the waters that excavated the valley, ascend this inclined plane to enter the cave G, nor would the external currents, however rapidly rushing by the outside of the mouth E, have power to agitate (except by slight eddies in the lower part of the throat F) the still waters that would fill the body of the cavern, and which being there quiescent, would, as at Kirkdale, deposit a sediment from the mud suspended in them upon the undisturbed remains of whatever kind that lay on the floor. From its low position, it is also probable that this vault formed the deepest recess of an extensive range of inhabited caves, to which successive generations of antediluvian bears withdrew themselves from the turbulent company of their fellows, as they felt sickness and death approaching; the habit of domesticated beasts and birds to retire and hide themselves on the approach of death, renders it probable that wild and savage animals also do the same. The unusual state of decay of the teeth and bones in this black earth may be attributed to the exposed state of this cavern, arising from its large mouth and proximity to the external atmosphere, and to the absence of that protection which in closer and deeper caves they have received, by being secluded from such exposure, or embedded in more argillaceous earth, or invested with and entirely sealed up beneath a crust of stalagmite.

GENERAL REMARKS ON THE GERMAN CAVES.

To the facts already enumerated in describing the particular appearances in the interior of individual caverns may be added the following remarks, which apply generally to them all.

1. With respect to the apertures themselves, whether fissures or caverns, they appear to have been open, and without mud or pebbles, at the time when the animals lived and died, whose remains are now found in them. These two kinds of apertures rarely occur separate, and many of the caves appear to be only enlargements, and hollow side branches shooting off from a fissure or congeries of connected fissures. Some of these fissures terminate upwards, like the caves, in the body of the rock; others rise to the surface, and are occasionally open, as at Scharzfeld, but more frequently they are filled up with diluvium of the same kind as that which has universally covered the floor, and filled the undervaultings of the caves.

2. The present mouths of the caves did not exist in their actual state at the time when they were inhabited; but are rather truncated portions of the lower regions of the original caverns, laid open by the diluvial waters which excavated the valleys in whose cliffs they stand; and which also drifted into them the mud and pebbles.

3. The proportion of teeth in all these caverns does not appear (as at Kirkdale) to be in excess, beyond that which is due to the number of bones that accompany them; and this circumstance is explained by the fact of their being principally derived from bears,

whose habit it is not to devour the bones, either of their own species, or of other large animals.

4. The partial occurrence of these remains in so comparatively small a number of the many caves that lie adjacent to each other, added to the immense quantities in which they are usually crowded together where they exist at all, shows that they were accumulated by some cause independent of the diluvial action that introduced the mud and pebbles; for had they been drifted in together, they would probably have been distributed co-extensively with these latter substances, and in small quantities; whereas, on the contrary, whilst we find in every cave nearly the same proportion of diluvial loam and pebbles, the occurrence of bones is limited to a small number; and in these, they are crowded in such enormous quantities, and are attended with such circumstances, as are explicable only on the hypothesis of their having existed there before the introduction of the diluvium; and in general, the deeper we descend, the more abundantly loaded do we find the lower regions and undervaultings to be, till they are entirely choked up with mud, pebbles, and bones.

5. The mud and pebbles were not introduced at a period anterior to that in which the caves were inhabited; for in this case, they would have found a separate bed at the bottom, beneath the bones, and not have been dispersed so equally as they are amongst them: e. g. we find the pebbles occur as abundantly at к the top, as at к. к. the bottom, and at L, the middle region of the great heap that lies piled together to the height of 25 feet in the lowest region of the cave, at Gailenreuth. (See Plate XVII.)

6. The angular fragments of limestone that are found within the

caves may have fallen from the roof either before or since the introduction of the diluvium, according as they are placed below or above
the stalagmitic crust that covers its surface; some of them below it
may also have been drifted in together with the pebbles.

7. With respect to the stalagmite, though it often occurs trans-
fused bodily through the substance of the diluvial sediment, it is
never found in continuous strata alternating with other strata of mud,
or pebbles, but always forming a single crust on the upper surface of
the sediment.   I could not find in any of the German caves a lower
crust of stalagmite formed as at Kirkdale, beneath the mud, on the
surface of the subjacent limestone rock; but from the thickness of
the diluvium, there were so very few points in which it was possible
to make any observations on this subject, that at present we are
without any evidence as to its existence or otherwise.

8. The diluvium itself is either simply a mass of pebbles, or of
loam or sand, or (which is more common) an irregular admixture of
all these three substances, having bones indiscriminately distributed
throughout them all; and in proportion as the mass has been more
or less percolated by stalagmitic infiltrations, the bones are either
simply embedded in loose earth, or in semi-indurated loam and
pebbles, or cemented together with loam and pebbles into a firm
osseous breccia, resembling that found in the fissures of limestone at
Gibraltar, and along the shores and islands of the Mediterranean.
Should it be suggested, that this loam or earthy matter may have
originated from dust that has fallen from the decomposition of the
roof of the caverns, the improbability of this origin appears from its
non-agreement in chemical composition with the limestones of these

roofs; whilst on the other hand, its perfect agreement with the diluvial loam that abounds on the surface of the adjacent countries, added to the fact of the materials within the cave being often sorted, or drifted, as if by water into distinct deposits of loam, and sand, and pebbles; and the still more irresistible argument, arising from the almost universal presence of the pebbles themselves, renders it impossible to refer the earthy matter in question to any cause acting exclusively within the interior of the caverns, or indeed to any other origin, than one violent movement of waters over the land without : the effects likely to have been produced by such an inundation on the interior of caverns having communication with the then existing surface, are precisely such as we find to have actually taken place, and to be attended by circumstances, all of which are consistent with the hypothesis of the mud and pebbles having been superinduced upon bones already existing in the caverns, by the waters of a transient deluge.

The facts I have enumerated in the above descriptions go to establish a perfect analogy, as far as relates to the loam and pebbles and stalagmitic incrustations in the caves and fissures of Germany and England, and lead us to infer an identity in the time and manner in which these earthy deposits were introduced; and this identity is still further confirmed by the agreement in species, of the animals whose remains we find enveloped by them, both in caves and fissures, as well as in the superficial deposits of similar loam and pebbles on the surface of the adjacent countries; viz. by the agreement of the animals of the English caves and fissures, not only with each other, but also with those of the diluvial gravel of England, and of the

U

greater part of Europe: and in the case of the German caves, by the identity of their extinct bear with that found in the diluvial gravel of Upper Austria; and of the extinct hyæna with that of the gravel at Canstadt, in the valley of the Necker; at Horden, near Herzberg, in the Hartz; at Eichstadt, in Bavaria; the Val d'Arno, in Italy; and Lawford, in Warwickshire. To these may be added the extinct rhinoceros, elephant, and hippopotamus, which are common to gravel beds as well as caves; and hence it follows, that the period at which the earth was inhabited by all the animals in question was that immediately antecedent to the formation of those superficial and almost universal deposits of loam and gravel, which it seems impossible to account for unless we ascribe them to a transient deluge, affecting universally, simultaneously, and at no very distant period, the entire surface of our planet *.

* I have much pleasure in referring my readers to two short and excellent papers in the Philosophical Transactions for 1794, Part II., by the Margrave of Anspach and the ever-memorable John Hunter; the former describing the caves of Gailenreuth, and the latter their organic remains, in a manner which cannot fail to be highly interesting to those who have followed me in my present description of them.

The account given by his Serene Highness is accurate and spirited; and it had not escaped him that the stalagmitic crust of the floor did not reach down to the bottom of the cave, but that there was a collection of what he calls animal rubbish between it and the actual floor of solid rock; but he overlooks the existence of pebbles, and adopts the two common errors of considering the diluvial loam as animal earth, and stating that the bones are found sticking every where in the sides of the cave, as well as lying on the bottom. I have more than once explained the source of this mistake, and pointed out the limitation within which the assertion is to be received.

Mr. Hunter accompanies his description of the bones with some curious speculations on the revolutions which may have affected the earth's surface, and some general observations on the different state of preservation of fossil animal remains; and concludes, as Cuvier has done since, with regard to the bones in the cave of Gailenreuth and others, that they have been accumulated in consequence of the cavities having been occupied as a place of retreat, or den of wild beasts during a long series of years—his words are, many thousand years; but he overrates this period considerably, grounding his opinion on the single fact of the different degrees of preservation of the bones in the same cavern. These, however, are not greater than those which exist in churchyard bones, or a common charnel-house, and may have been produced by the difference of a very few years, or certainly of a few hundred years, in the time of their exposure to decomposing causes at the bottom of the cave, before the introduction of the diluvial loam, which has since buried and protected them from any further considerable decay. Mr. Hunter's reasoning would have been correct, had there not existed this difference in the degree of exposure of the bones before and since the introduction of the loam: had he been aware of this fact, he certainly would have seen the force of it, and his expression would probably have been, many hundred years, instead of many thousand. I refer to my note on the cave of Kühloch, and to my account of Kirkdale, for further grounds on which I have founded my opinion as to the chronological inferences to be derived from the quantity of animal remains accumulated in these caves, and from the state and relative position of their stalagmite and diluvial loam and pebbles.

OSSEOUS BRECCIA OF GIBRALTAR, NICE, DALMATIA, &c.

The close connexion, or rather the identity of circumstances which we have seen to exist with respect to the time and manner in which the bones were introduced to both caverns and fissures, in Germany and England, leads me to think it almost certain, that those fissures also which are found to contain such large quantities of osseous breccia in the limestone rocks of Gibraltar, Antibes, Nice, Cette, Pisa, and Dalmatia, and numerous other places along the north shores of the Mediterranean and Adriatic, and in the islands of Cerigo, Corsica, Sardinia, Sicily, &c., had become charged with these remains in the same antediluvian period with the caves and fissures I have been describing. M. Fortis, in his account of the breccia of Dalmatia, and some of the islands, says it occurs both in vertical and horizontal cavities of the limestone, and that it is visible in clefts and fissures along the shores, and in caves in the interior of all the islands and coasts of Illyria; that the bones are usually embedded in a red ochreous cement, dispersed and broken, and that a single skeleton is never found entire. M. Provençal repeats the same observations in his account of the breccia in the caves and fissures near Nice. M. Chevalier also says of the bones at Gibraltar, that they lie separate one from another, but not rolled, and that the greater part of them appear to have been broken before they were incrusted in their present cement: and M. Cuvier, in his first edition, has given a list of the animals of this breccia; among which he enumerates the ox, deer, antelope, sheep, rabbits, water-rats, mice, horse, ass, snakes, birds, and land-shells. He states, also, that the greater number of them decidedly agree with existing species, and supposes them to have fallen into the fissures in the period succeeding the last retreat of the waters. With respect to some of these bones, it is probable that this hypothesis is correct, and that here, as well as in England, there may exist, in addition to the breccia containing bones of antediluvian origin, other more recent deposits, derived from animals which are continually falling into the comparatively few fissures which are still open, as at Duncombe Park; but with regard to others, viz. to those which occur in fissures that are closed up, as at Gibraltar, to the very surface of the soil, the case is different, and their origin clearly antediluvian.

In my former edition I had ventured to differ from M. Cuvier on this subject, judging from the apparent agreement in species of many of the graminivorous animals of the osseous breccia with those found in the antediluvian cave at Kirkdale, and in the diluvial gravel-beds of England, and had suggested that the discovery in the Mediterranean breccia of any of the extinct species of animals we find in the caves, or in diluvian gravel, would establish the higher antiquity I am contending for. Being at Paris in October last, I had the satisfaction to be informed by M. Cuvier, that he has lately found the tusks of the extinct lion or tiger in the breccia of Nice, and that he has added other animals belonging to species now unknown, to the list he had before given. Mr. Pentland also has recently discovered the tooth of the same extinct tiger in the breccia of Antibes, and has found in the cabinet of Professor Targioni, at Florence, the femur of a bear from the osseous breccia of Pisa, and has been informed that other similar bones occur in the neighbourhood of Sienna.   I was still further gratified by M. Cuvier's showing me specimens from several of the places above enumerated, many of which contained rolled pebbles, and all of them a large proportion of indurated earthy loam; through which, as their matrix, the teeth and fragments of bones are disseminated in a manner no way different from that in which they occur in the indurated loam at Plymouth and in the caves of Germany.   This loam is described in many accounts of the osseous breccia, as being ochreous stalactite, but this description is incorrect; it is a mass of earthy loam, differing only in colour from that which fills the caves and fissures, and composes the superficial diluvial loam in Germany; and its consolidated state arises from the stalagmitic infiltrations that have percolated its pores, and formed thin veins and linings of calc sinter in the innumerable crevices and small cellular cavities with which it is interspersed.   This is precisely the state of much of the loam in the caves of Germany; and in both cases the admixture of pebbles with angular fragments of limestone, and the irregular manner in which the bones, though evidently not rolled, are broken and crowded together in confused heaps, seem to indicate that, as I have suggested in my explanation of the bones and breccia at Plymouth, they have been moved within the cave or fissure by water, to a small distance only from the spot where they fell in and died, and simply broken by this removal, but neither rounded or reduced to pebbles.   The same waters which would thus drift them into irregular heaps in the bottom of the caves or fissures, may also have introduced the loam and pebbles through which they are in each case dispersed. The chief differences with regard to the Mediterranean breccia seem to consist, 1st. in the loam being red, instead of its more usual colour, brown; an accident which may be explained by the hypothesis of its being diluvial detritus, derived from strata of a red colour, and which is rendered probable by the fact mentioned by Major Imrie, of there being, on the summit of the rock of Gibraltar, superficial deposits of a similar red earth, which from his description are clearly diluvial. 2d. In the proportion of angular fragments of stone being greater in this breccia than in that from the caverns: this may be referred to their having fallen in greater abundance from the sides of the fissures (whilst they were yet open), than from the roof of close caverns, in consequence of the greater exposure of the former to the decomposing influence of the atmosphere; and this hypothesis is corroborated by the fact spoken of by Mr. Allan, in his excellent Paper on the Geology of Nice (Edin. Phil. Trans. vol. viii. part 2), that the naked surfaces of the limestone rocks near Nice (which are of the same kind with that of Gibraltar), are broken and shattered into angular fragments, which lie on the surface of the mountains, and are mentioned by Saussure under the name of " bréche en place *." 3d. A third point wherein the breccia we are considering differs from that of the caverns is, that it contains land-shells. These may be considered as having fallen in from the sides of the fissures, together with the animals and the angular fragments last spoken of; whilst the depth and covered state of the

---

* My own observation has presented to me many similar occurrences of loose naked angular fragments on the surface of many other compact limestone rocks.

caves would allow no such circumstance to have occurred in them. 4th. A fourth difference arises from the remains being chiefly those of graminivorous animals; and this is consistent with the circumstance I mentioned when speaking of the recent bones in Duncombe Park, that such animals are more liable than beasts of prey to fall into open fissures, from their constant habit of traversing the surface in the act of cropping the grass which forms their food.   All these differences may be explained on the theory I am maintaining, that the bones in the osseous breccia are of antediluvian origin, and coeval with the remains we find in the caves and fissures of Germany and England.

M. De Luc has expressed precisely the same opinion at the conclusion of his description of the bones, earth, and breccia contained in the caverns of the Hartz.   " L'aspect de la couche d'ou l'on tire ces os, ne permet pas de douter de leur origine; elle est la même que celle des os de Dalmatie et de Gibralter, ainsi que de tous les autres corps terrestres ensevelis dans les couches de nos continens." By which latter he means the bones of elephant, rhinoceros, &c. that occur so universally in the diluvium.—De Luc's Lettres Phys. et Mor. vol. iv. p. 90.

Mr. Allan, in his paper just quoted, designates the earthy matrix of these bones as " Red indurated clay;" and adds also, that "with the bones are rounded pebbles of limestone." He also states his opinion, that the bones have been deposited in two distinct eras, which, I have no doubt, will on explanation turn out to be the same antediluvian and postdiluvian periods to which I have assigned respectively the introduction of different bones into the fissures of Plymouth and Duncombe Park.   "On the castle rock at Nice," says he, " the bones occur in two distinct states, one forming a very hard indurated breccia, the cement of which varies from a brown colour to almost black; in the other, they are loose, or feebly agglutinated, by means of calcareous infiltrations, with fragments of limestone and sea shells*." He adds, there appear to be several fissures, some containing a few dispersed fragments of bone, and others of loose earth and stones.

All these arguments are still further corroborated by the state of preservation of the Gibraltar bones, being exactly that of the bears bones, which occur in the osseous breccia of the caves of Gailenreuth, &c.; they are white, dry, and adherent to the tongue, different from and indicating a much higher antiquity than the postdiluvian bones that occur above the stalagmite crust within the German caves, or in the open fissure at Duncombe Park.

But the fact of the breccia of Gibraltar (and by consequence those of Nice, Cette, Dalmatia, and the other places before enumerated) being coeval with that of the caverns we have been describing, is, I think, established beyond all doubt by the minute and careful account we have of it in Major Imrie's Mineralogical Description of Gibraltar, in the 4th vol. of the Transactions of the Royal Society of Edinburgh, which is so important, that I shall here extract those parts of it which bear on the point before us:—"The insulated mountain, or rock of Gibraltar," says he, "is composed of compact limestone, rising at its greatest elevation 1439 feet above the level of the sea, being about three miles long, and three quarters of a mile broad in its widest part,

* This occurrence of marine shells in the looser variety of breccia may possibly be attributed to their having been collected by sea birds or by men, as in the case of the cave of Paviland.

and bounded for the most part by rugged slopes, or by precipices ; like other compact limestones, it is perforated by caverns of vast extent, and also by vertical fissures.   The largest cavern (St. Michael's) is 1000 feet above the sea, and consists of a long series of caves, of difficult access, which have been penetrated to the distance of 300 feet from the first cavern, and extend still further, and abound with stalactites.   In this cave no bones have yet been noticed, but in the perpendicular fissures of the rock, and in some of the caverns (all of which afford evident proofs of their former communication with the surface) a calcareous concretion is found of a reddish brown fer-ruginous colour, with an earthy fracture and considerable induration, inclosing the bones of various animals ; these bones are of various sizes, and lie in all directions, intermixed with shells of land snails, fragments of the calcareous rock, and particles of spar.   Bones com-bined in similar concretions are found also in Dalmatia, the islands of Cherso, and Ossero, and have been described by Fortis ; and by his account it appears, that with regard to situation, composition and colour, they are perfectly similar to those found at Gibraltar, and occur in fissures and caves of the same species of limestone.   I have traced (says Major Imrie) in Gibraltar this concretion from the lowest part of a deep perpendicular fissure up to the surface of the mountain ; as it approached the surface, it became less firmly combined, and when it had no covering of the calcareous rock, a small degree of adhesion only remained.

" At Rosia Bay, on the west-side of Gibraltar, this concretion is found in what has evidently been a cavern originally formed by huge unshapely masses of rock, which have tumbled in together.   The fissure, or cavern, formed by the disruption and subsidence of these masses, has been evidently filled up with this concretion, and is now exposed to full view, by the outward mass having dropped down, in consequence of the encroachments of the sea. It is to this spot that strangers are generally led to examine the phenomenon; and the composition having here attained to its greatest degree of hardness and solidity, the hasty observer seeing the bones inclosed in what has so little the appearance of having been a vacuity, examines no further, but immediately adopts the idea, of their being incased in solid rock. The communication from the former chasm, to the surface from which it has received the materials of the concretion, is still to be traced in the face of the rock, but its opening is, at present, covered by the base of the line-wall of the garrison. Here bones, apparently human*, are scattered among others of various kinds and sizes, even down to the smallest bones of birds.

" At the north extremity of the mountain, the concretion is generally found in perpendicular fissures; the miners there employed upon the fortifications, in excavating one of those fissures, found at a great depth from the surface two skulls (not human.) This concretion varies in its composition, according to the situation where it is found. At the extremity of Princes Lines, high in the rock which looks towards Spain, it is found to consist only of a reddish calcareous earth, and the bones of small birds, cemented thereby. The rock around this spot is inhabited by a number of hawks, that in the breeding season nestle here and rear their young; the bones in this

* This error, as to the existence of human bones in this breccia, has been since corrected.

concretion are probably the remains of the food of those birds.   At the base of the rock below Kings Lines, the concretion consists of pebbles of the prevailing calcareous rock.   In this concretion, at a considerable depth under the surface, was found part of a green glass bottle."

The above extracts need but little comment.   The birds' bones at the rock of Princes Lines, and the glass bottle buried under pebbles at the base of Kings Lines, show that there is still going on daily a formation of postdiluvian breccia; whilst the whole description of the earthy contents of the caves and fissures, and of the manner in which they are filled, is so entirely like that I have given of the caves and fissures of Germany and England, that it seems to me impossible to hesitate in admitting the identity of their origin.   But the proof (satisfactory as it is) does not stop here; I would attribute, in this case, as in the caves at Plymouth, the introduction of the loam and pebbles to diluvial action superinduced upon bones and angular fragments, that had fallen into the cavities whilst yet open, in the period preceding the last general inundation of the earth; and in the paper I have just quoted, Major Imrie has supplied one of the most neat and convincing proofs I ever met with, that this same diluvial action has been exerted on the summit of the very mountain whose fissures I am contending have been filled by it.   Describing the upper surface of this mountain, he says, " The uncovered parts of the rock expose to the eye a phenomenon worthy of some attention, as it tends clearly to demonstrate, that however high the surface of this rock may now be elevated above the level of the sea, it has once been the bed of agitated waters.   This phenomenon is to be observed in many parts of the rock; it consists of pot-like holes of various sizes, hollowed out of the solid rock, and formed apparently by the attrition of gravel or pebbles, set in motion by the rapidity of rivers or currents in the sea. One of those which had recently been laid open I examined with attention, and found it to be five feet deep, and three in diameter; the edge of its mouth rounded off, as if by art, and its sides and bottom retaining a considerable degree of polish. From its mouth, for three and a half feet down, it was filled with a red argillaceous earth, thinly mixed with minute parts of transparent quartz crystals; the remaining foot and a half, to the bottom, contained an aggregate of water-worn stones, which were from the size of a goose's egg to that of a small walnut, and consisted of red jaspers, yellowish white flints, white quartz, and bluish white agates, firmly combined by a yellowish brown stalactitical calcareous spar. In this breccia I could not discover any fragment of the mountain rock, or any other calcareous matter, except the cement with which it was combined. This pot is 940 feet above the level of the sea."

Now, comparing these facts with the phenomena he had before described, we see that the red earth here mentioned is the very substance which, in the caves and fissures immediately below, forms the diluvial matrix in which the bones are embedded, and together with which they have been united by stalagmitic infiltrations into a mass of solid osseous breccia; and the pebbles of quartz, agate, jasper, &c. lodged with this red earth on the summit of an insulated, lofty, and precipitous mountain of naked limestone, present a case analagous to the blocks of Mont Blanc granite on the limestone mountains of the Jura; both being on spots from which it is impossible they could have derived their origin, and to which they could have been trans-ported by no other force than that of a tremendous deluge, or de-bacle of water drifting them from a great distance to the place they at present occupy; and in which, like all the other deposits of this grand catastrophe, they have remained ever since undisturbed, on the very spot on which they were cast at the time of the last great geo-logical change, by an inundation of water that has affected universally the surface of the earth.

To the above extracts from Major Imrie, I am enabled to add the testimony of another gentleman, now resident at Gibraltar, which has just been forwarded to me from thence by my friend the Rev. R. Curtois. As the information comes in the form of a letter to Mr. Curtois from Mr. C. Pargeter, a medical officer at Gibraltar, and who proposes himself to publish an account of the bones they find in that spot, I will subjoin an extract from it, which is equally corroborative of my views with the account I have just transcribed from Major Imrie.

" The bones," says he, " are found imbedded, 1st. in an ochreous sandy earth, cemented by calcareous matter, and much indurated, together with angular fragments of limestone; or, 2nd. in a mass, which may be termed pudding-stone; consisting of pebbles of white quartz, and of variously coloured flinty pebbles (of the same nature as noticed by Colonel Imrie, as occurring in the pot-holes at the summit of the rock) and of limestone; all of them much rounded, and varying in size from minute gravel to that of a goose's egg; ma-rine shells, but in small number, are sometimes found with the bones in this mass, and all the materials are firmly cemented together; or,

3dly, the bones occur in a mass, composed of rounded pebbles and reddish earth. These three varieties are to be seen together, both on the western and eastern side of the rock, particularly the latter; the bones, sand, pebbles, &c. are found at various heights above the sea, from a few feet to five and six hundred." He then subjoins the following sketch of the position of the breccia on the floor of some caves facing the sea, which I have here copied, as being almost a fac simile of the section of the two caves facing the sea at Paviland, near Swansea.

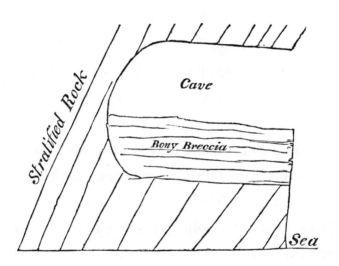

Mr. Pargeter adds further, " It has not been ascertained, that there are here any other varieties of bones than those described by M. Cuvier; we have no remains of elephant, rhinoceros, hippopotamus, &c., though the two first have been lately found in an alluvial calcareous rock, at Tarifa, and the same alluvial formation extends for a great way round the Bay of Gibraltar." Mr. Curtois also states, that among the remains discovered at Tarifa, is the tusk of an elephant, the chord of whose curve is six feet, and that the greatest quantity of osseous breccia is in a cave opposite Ceuta*.

All these circumstances concur to establish, as far as any evidence short of personal examination can establish, an identity of time in the formation of the osseous breccia, in the fissures and caves of Gibraltar, and the coast and islands of the Mediterranean, with that of the bones which occur in the caves and fissures of Germany and England; and to show, that in each case, the period in which the animal remains were introduced to them was that immediately preceding the inundation, which superadded the mud and pebbles in which they are now enveloped.

In the adjacent country of Spain, Mr. Bowles has described some caves at Concud, near Teruel, in Arragon, in a rock of Shelly limestone, in which they find bones and teeth of ox, horse, ass, sheep, and other animals; some solid, and in the state of common grave bones, others calcined and falling to powder; and also human bones, of which the cavities are full of crystalline matter. These in some cases lie in loose earth with rolled pebbles, in others they are united into blocks of hard rock, from four to eight feet long. The details of Mr. Bowles's description are given by M. Cuvier, and though they are indistinct, it yet seems probable, that here also as at Gibraltar, and

---

* The extensive alluvial calcareous rock here spoken of is the same which I have called diluvium, and its consolidated state arises from the abundance of calcareous matter which pervades it. A similar consolidation occurs in the calcareous diluvium I have before mentioned as containing the bones of bears at Kremsminster, in Upper Austria; and is very common in all countries, in the case of gravel beds that contain a large proportion of calcareous pebbles, or calcareous sand.

near Nice, we have osseous deposits of two eras, one recent and post-diluvian, the other immediately antediluvian.

In Italy, M. Cuvier mentions several caverns containing similar deposits of bones, to which I am enabled to add, on the authority of Mr. Pentland, that the bones of ruminating animals have been found united by stalagmite to a breccia, like that of Nice and Antibes, in the Grotto Della Molpa, at Cape Palinurus, in the kingdom of Naples; and that in the Sabine Mountains, not far from Tivoli, near the town of Palombaro, there is a cave in which the remains of the bear have been found incrusted with stalagmite, and of which there is a jaw in the collection of Professor Canali, of Perugia.

In the gorge, at Peckaw, in Styria, by which the Mur runs down from Brück to Gratz, there is a lofty vertical cliff of Alpine limestone, in the perpendicular face of which I observed the mouths of several caverns; some of these have been found to contain also the bones of the great extinct species of bear. The position of these caves with respect to this diluvian gorge is analogous to that of the caves in the cliffs that flank the valley of Muggendorf, in Franconia, and of the Avon, at Clifton.

An account has recently been published of bones of the same kind of bear discovered at Adelsberg, in Carniola. And a breccia, like that of Gailenreuth, and containing also the bones of bears, has been found in the mountain of Ischuber, in Croatia.

For a detailed account of further localities of these ossiferous caves and fissures, extending as they do over England, Spain, France, Italy, Dalmatia, Croatia, Carniola, Styria, Austria, Hungary, Poland,

Y

and Germany, I must refer my readers to the 'Ossemens Fossiles of Cuvier;' a work containing more sound and philosophical reasoning on the early state of our planet, and a more valuable collection of authentic facts relating to the history of its fossil animals of the higher orders, than can be found in all the books that have ever yet been written upon the subject.

In the conclusion of my account of Kirkdale, I stated, that its phenomena were decisive in establishing the fact, that animals which are now limited exclusively to warmer latitudes, e. g. the elephant, rhinoceros, hippopotamus, and hyæna, were the antediluvian inhabitants of Britain, and not drifted northwards by the diluvian currents from more southern or equatorial regions, as had often been suggested, and was never till now disproved; and I pointed out the inference with respect to a probable change of climate in the northern hemisphere, which seems to follow from this circumstance.

Another important consequence arising directly from the inhabited caves, and ossiferous fissures, the existence of which has been now shown to extend generally over Europe, is, that the present sea and land have not changed place; but that the antediluvian surface of at least a large portion of the northern hemisphere was the same with the present; since those tracts of dry land in which we find the ossiferous caves and fissures must have been dry also, when the land animals inhabited or fell into them, in the period immediately preceding the inundation by which they were extirpated. And hence it follows, that wherever such caves and fissures occur, i. e. in the greater part of Europe, and in whatever districts of the other Continents such bones may be found under similar circumstances, there did not take place any such interchange of the surfaces occupied respectively by land and water, as many writers of high authority have conceived to have immediately succeeded the last great geological revolution, by an universal and transient inundation which has affected the planet we inhabit.

HUMAN REMAINS IN CAVES.

It was mentioned, when speaking of Gailenreuth and Zahnloch, that human remains and urns had been discovered there in the same cave with the bones of antediluvian animals, but that they are of comparatively low antiquity.

Six analogous cases have been noticed in this country in cavities of mountain limestone, in the counties of Somerset, Glamorgan, Caermarthen, and York; and these also are attended by circumstances which indicate them to be of postdiluvian origin.

1. The discovery of human bones incrusted with stalactite, in a cave of mountain limestone at Burringdon, in the Mendip-hills, and mentioned in Collinson's History of Somerset, is explained by this cave having either been used as a place of sepulture in early times, or been resorted to for refuge by wretches that perished in it, when the country was suffering under one of the numerous military operations which, in different periods of our early history, have been conducted in that quarter. The mouth of this cave was nearly closed by stalactite, and many of the bones incrusted with it. In the instance of a skull, this substance had covered the inside as well as the outside of the bones; and I have a fragment from the inside, which bears in relief casts of the channel of the veins along the interior of the skull. The state of these bones affords indications of very high antiquity; but there is no reason for not considering them

postdiluvian.   Mr. Skinner, on examination of this cave, found the
bones disposed chiefly in a recess on one side, as in a sepulchral catacomb; and in the same neighbourhood, at Wellow, there is a large
artificial catacomb of high antiquity, covered by a barrow, and constructed after the manner of that at New Grange, near Slane, in the
county of Meath, of stones successively overlapping each other till
they meet in the roof.   In this were found the remains of many
human bodies.   A description of it may be seen in the Archæologia
for 1820.

2. Mr. Miller, of Bristol, has lately discovered the remains of human
bodies in the much frequented cave of Wokey Hole, near Wells, at
the south-west base of the Mendips.   On hearing of the fact in
January last, I went the next day to examine it, and found the bones
to be placed in the most secluded and distant part of a large fissure
that shoots off laterally from this cave, and is separated from its main
chambers by a subterraneous river of considerable size, that constantly
runs through them.   They have been broken by repeated digging to
small pieces; but the presence of numerous teeth establishes the
fact that they are human.   These teeth and fragments are dispersed
through reddish mud and clay, and some of them united with it by
stalagmite into a firm osseous breccia.   Among the loose bones I
found a small piece of a coarse sepulchral urn.   The spot on which
they lie is within reach of the highest floods of the adjacent river,
and the mud in which they are buried is evidently fluviatile, and not
diluvian; so also is great part, if not the whole, of the mud and
sand in the adjacent large caverns, the bottoms of all which are filled
with water to the height of many feet, by occasional land-floods,

which must long ago have undermined and removed any diluvial deposits that may have originally been left in them. I could find no pebbles, nor traces of any other than the human bones, on the single spot I have just described; these are very old, but not antediluvian. In another cave on this same flank of the Mendips, at Compton Bishop, near Uxbridge, Mr. Peter Fry, of Axbridge, discovered in the year 1820 a number of bones of foxes, all lying together in the same spot, and brought away 15 skulls. These also, like the remains of foxes in Duncombe Park and near Paviland, are of postdiluvian origin, and were probably derived from animals that retired to die there, as the antediluvian bears did in the caves of Germany.

3. Mr. Dillwyn has observed two analogous cases in the mountain limestone of South-Wales; one of these was discovered, in 1805, near Swansea, in a quarry of limestone at the Mumbles, where the workmen cut across a wedge-shaped fissure, diminishing downwards, and filled with loose rubbish, composed of fragments of the adjacent limestone, mixed with mould. In this loose breccia lay, confusedly, a large number of human bones, that appear to be the remains of bodies thrown in after a battle, like those I have mentioned near Kirby Moorside in Yorkshire, with no indications of regular burial; they were about 30 feet below the present upper surface of the lime-stone rock.

4. The other case occurred in 1810 at Llandebie, in Caermarthen-shire, where a square cave was suddenly broken into, in working a quarry of solid mountain limestone on the north border of the great coal basin. In this cave lay about a dozen human skeletons in two rows at right angles to each other. The passage leading to

this cave had been entirely closed up with stones for the purpose of concealment, and its mouth was completely grown over with grass.

5 and 6. For the particulars of these two cases I refer my readers to the descriptions I have already given of the human remains in the cave of Paviland, in Glamorganshire, and in the fissure at the back of the parks, near Kirby Moorside, in Yorkshire.

It is obvious, that in none of these cases are the bones referable to so high an era as those of the wild beasts that occur in the caves at Kirkdale, and elsewhere.

The 7th and last case I shall mention, is one which has recently been published, with a very able and judicious commentary by Mr. Weaver, in the Annals of Philosophy for January 1823; wherein he gives a translation of Baron Schlotheim's account of human bones discovered in the Valley of the Elster, near Köstritz in Saxony, a few leagues south-west of Leipsig, and of which I shall subjoin the following abstract. " The Valley of the Elster, near Köstritz, is flanked by hills, the summits of which are composed of limestone, locally called Zechstein, whilst the lower regions contain beds and large masses of gypsum. Both limestone and gypsum contain caves and fissures, which are in each case equally filled with a mass of loam or clay of the same kind as that which covers the adjacent country. In this loam are various pebbles of limestone, and of rocks that occur only at a distance, e. g. granite, &c. The principal deposits of bone are in the loam which fills the cavities of the limestone : among these there occur at Politz the remains of rhinoceros, horse, ox, hyæna, tiger (or jaguar), and bear; they are in the same state as the bones at Gailenreuth and Scharzfeld, and probably of the same era.

The cavities of the gypsum are very numerous, communicating with each other, and traversing the rock in all directions : in the loam which fills them the bones are dispersed irregularly in clusters, or collected in heaps, without order, and at different depths, and have been continually discovered from the first opening of these quarries, thirty years ago, to the present time.   They consist of the following animals :

    No. 1. Rhinoceros, deer, ox, jaguar, and hyæna, the same as in the limestone cavities, excepting the remains of horse.

    No. 2. Sheep or roe, fox, weasel, squirrel, field-mouse, shrew-mouse, common rat, hamster rat, bat, mole, (five portions of the jaw of young ones), hare, rabbit, bat, frog, two species of owl, domestic-cock, and man.

These bones, No. 2, occur mixed confusedly, not only with one another, but also with the bones of the extinct animals; they all belong to existing species, and are in various stages of decay, but are less calcined than the bones of extinct animals, No. 1.   Remains similar to them are found also in the soil of the adjacent fields.   In one quarry (called Winters), the human bones were found eight feet below those of rhinoceros, and 26 feet below the surface.   It is highly probable, as M. Schlotheim himself suggests, from the admixture of the bones of so many species of recent animals with the human remains in the gypsum quarries, that both these are of later origin than those in the limestone; they appear, I think, to have been introduced at a subsequent period into the diluvian loam, which had before contained the more ancient bones and pebbles; but by what means, or at what precise period of the postdiluvian era, remains yet to be ascertained.

M. Schlotheim says, " I am far from thinking satisfactory the explanation I have attempted of these phenomena, and am disposed to consider the human bones to be of a later epoch than the larger land animals of the ancient world; all other reported cases of human remains accompanying the bones of beasts of prey have not been confirmed on closer investigation." He dwells also on the circumstance, that the limestone cavities which are situated in the hills contain only the remains of ancient animals, whilst ancient and modern bones occur mixed together only in the gypsum cavities, the position of which is at a lower level in a kind of basin at one of the lowest parts of the district.

M. Schlotheim's hypothesis is that the ancient bones were washed out of the upper caves into the lower ones, and thus mixed with the modern bones, by a succession of floods, produced by the successive bursting of lakes in a higher part of the country; I fully agree with him in thinking this explanation unsatisfactory. The chief point, however, is conceded, viz. that the human bones are not of the same antiquity with those of the antediluvian animals that occur in the same caves with them; and thus far the case of Köstritz affords no exception to the general fact, that human bones have not been discovered in any of those diluvial deposits which have hitherto been examined; and in which, from the great abundance they contain of the remains of wild animals, that could not have existed in numbers sufficient to supply these remains, in a country inha-

z

bited by man, it is highly improbable that they ever will be found. On this important subject I fully coincide with the opinions expressed by Mr. Weaver, " that the satisfactory solution of the general problem, as far as it relates to man, is probably to be sought more particularly in the Asiatic regions, the cradle of the human race ; and that another interesting branch of inquiry connected with it is, whether any fossil remains of elephant, rhinoceros, hippopotamus, and hyæna, exist in the diluvium of tropical climates ; and if they do, whether they agree with the recent species of these genera, or with those extinct species, whose remains are dispersed so largely over the temperate and frigid zones of the northern hemisphere *."

* One probable reason why such remains have not been noticed in the banks of the rivers of Central and Southern Asia, and of Africa, may be, that in warm climates, they cannot have been preserved in ice as in the higher latitudes in a state of perfection fit for the purposes of commerce; and consequently, can have afforded to the natives no motive to collect them for sale. The absence of roads also in these barbarous countries, and consequent non-existence of open gravel pits (in which such remains are for the most part found in Europe), is another cause, which helps to explain the total ignorance in which we have so long stood, and are likely to continue, as to the presence or absence of bones of any kind in the diluvium of Central Asia, and Africa.

PART II.

HAVING thus far fulfilled my original proposal to illustrate my account of Kirkdale, and the caves of England, by a comparative view of similar caverns and fissures on the Continent, I come now to the second part of my inquiry, viz. the evidence of diluvial action afforded by the accumulation on the earth's surface of loam and gravel, containing the remains of the same species of animals that we find in the caves and fissures, and by the form and structure of hills and valleys in all parts of the world.

EVIDENCE OF DILUVIAL ACTION FROM THE DISPERSION OF THE BONES OF ELEPHANTS, &c.

As the fossil elephant is more generally dispersed, and has been more frequently noticed than any one of the other animals we find with it in the diluvial detritus of which I am about to speak, and as it is peculiar to, and may be considered characteristic of deposits of this

z 2

era, I shall introduce my remarks on the evidence of diluvial action afforded by deposits of loam and gravel, with a short history of the remains of this animal, and of the extent to which they occur in England; and with respect to the Continent, shall simply refer to Cuvier for proofs of their dispersion over every country, and almost every valley in Europe and northern Asia, as well as in North America.

The fossil elephant differs entirely from any living species of that genus, but approaches more nearly to the Asiatic than to that of Africa. Blumenbach has distinguished it by the name of elephas primigenius, and Cuvier of elephant fossile. The term mammoth (animal of the earth) has been applied to it by the natives of Siberia, who imagined the bones to be those of some huge animal that lived at present like a mole beneath the surface of the earth.

It appears from the wonderful specimen that was found entire in the ice of Tungusia, that this species was clothed with coarse tufty wool of a reddish colour, interspersed with stiff black hair, unlike that of any known animal; that it had a long mane on its neck and back, and had its ears protected by tufts of hair, and was at least sixteen feet high.

The bones of elephants occurring in Britain had from very ancient times attracted attention, and are mentioned with wonder by the early historians (see Harrison's Introduction to Hollingshed's Chronicle, ch. v. p. 17 and 21; also Roger de Coggleshall, as quoted by Camden, Collinson's Hist. of Somerset, vol. i. p. 180, and Plott's Oxfordshire, p. 132 to 139); but their history was never fully understood till the recent investigations of Cuvier. The old and vulgar notion that they were gigantic bones of the human species is at once refuted by the smallest knowledge of anatomy.   The next idea, which long prevailed, and was considered satisfactory by the antiquaries of the last century, was, that they were the remains of elephants imported by the Roman armies.   This idea is also refuted; 1st, by the anatomical fact of their belonging to an extinct species of this genus; 2dly, by their being usually accompanied by the bones of rhinoceros and hippopotamus, animals which could never have been attached to Roman armies; 3dly, by their being found dispersed over Siberia and North America, in equal or even greater abundance than in those parts of Europe which were subjected to the Roman conquests.   The still later and more rational idea, that they were drifted northwards by the diluvian waters from tropical regions, must be abandoned on the authority of the evidence afforded by the den at Kirkdale; and it now remains only to admit, that they must have inhabited the countries in which their bones are found.

It was to be expected that the remains of elephant should be found in the diluvial gravel of Yorkshire, from the fact already established, that these animals inhabited the neighbourhood of Kirkdale, whilst its caverns were occupied by hyænas; and accordingly, the teeth and tusks of bones of elephants of prodigious size have been found in the diluvium at Robin Hood's Bay, near Whitby, at Scarborough, Bridlington, and several other places along the shore of Holderness.   As we proceed southwards, we continue to find them abundantly on the coast, and in the interior of Norfolk, Suffolk, and Essex.   The largest deposit of them is at Walton, near Harwich, where they lie at the water's edge, mixed with great numbers of the teeth, bones, and horns of elk, stag, ox, horse, and other diluvial animals. In the valley of the Thames they have been discovered at Sheppy, the Isle of Dogs, Lewisham, London, Brentford, Kew, Hurley Bottom, Wallingford, Dorchester, Abingdon, and Oxford; also at Norwich, Canterbury, and Chartham, near Rochester. On the south coast of England they occur at Lyme Regis and Charmouth (from the latter place Mr. De la Beche has lately obtained a tusk nine feet eight inches in length); also at Burton and Loders, near Bridport, and near Yeovil in Somerset. At Whitchurch, near Dorchester, they lie in gravel above the chalk, and are found in a similar position on Salisbury Plain: in the valley of the Avon, also at Box, and Newton near Bath, and in that of the Severn, at Gloucester; and at Rodborough, near Stroud. In the centre of England we have them at Trentham, in Staffordshire, at two places mentioned by Grew and Morton, in Northamptonshire, and at Newnham and Lawford, near Rugby, in Warwickshire. In North Wales, Pennant mentions two molar teeth and a tusk found in Flintshire, at Holkin, near the mouth of the Vale of Clewyd; and they are not wanting, though they have been less frequently noticed, in Scotland and Ireland. In all these cases they are found in the superficial diluvial detritus, consisting either of gravel, sand, loam, or clay, and are never embedded in any of the regular strata.

The circumstances that attend some of these deposits require to be more particularly detailed. In the streets of London the teeth and bones are often found, in digging foundations and sewers, in the gravel; e. g. elephants' teeth have been found under 12 feet of gravel in Gray's-Inn Lane; and lately at 30 feet deep, in digging the grand sewer, near Charles-street, on the east of Waterloo Place. At Kingsland, near Hoxton, in 1806, an entire elephant's skull was discovered, containing two tusks of enormous length, as well as the grinding-teeth: they have also been frequently found at Ilford, on the road from London to Harwich, and, indeed, in almost all the gravel-pits round London. The teeth are of all sizes, from the milk-teeth to those of the largest and most perfect growth; and some of them show all the intermediate and peculiar stages of change to which the teeth of modern elephants are subject. In the gravel-pits at Oxford and Abingdon, teeth and tusks, and various bones of the elephant, are found mixed with the bones of rhinoceros, horse, ox, hog, and several species of deer, often crowded together in the same pit, and seldom rolled or rubbed at the edges, although they have not been found united in entire skeletons*.

In the Ashmolean Museum there are some vertebræ, and a thigh-bone of an enormous elephant, at least sixteen feet high, which are in the most delicate state of preservation, and were found in the gravel at Abingdon four years ago. In the same pit with them they collected also fragments of sixteen horns of deer. These bones and horns are extremely soft and brittle whilst wet, but harden by drying: they are not in the smallest degree mineralised, but retain less of their animal matter than those which have been laid in clay or loam; they are very adherent to the tongue. About three years since

---

* For a further detail of the gravel beds of Oxford, Witham Hill, and Bagley Wood, and of the organic remains contained in them, I must again refer to chap. xvii. of Dr. Kidd's Geological Essays.

a large molar tooth of an elephant was dug up in a gravel-pit in one of the streets of Oxford, in front of St. John's College.

In the Philosophical Transactions for 1813, is a report of the tusk of an elephant, nine feet long, and of other remains of the same animal, with those of hippopotamus, ox, and several species of deer, and the horn of an ox, four feet and half in length, all of which were found by Mr. Trimmer in the gravel of the valley of the Thames, near Brentford.   Six tusks of the hippopotamus lay in an area of 120 yards.   At Plate XXII. fig. 5, I have copied from Lee's Natural History of Lancashire the entire head of an hippopotamus, found in that county under a peat bog.   In all these gravel beds rarely two bones lie in immediate contact with each other, and in very few cases are they rounded by attrition.

At Newnham, in Warwickshire, near Church Lawford, about two miles west of Rugby, two magnificent heads and other bones of the Siberian rhinoceros, and many large tusks and teeth of elephants, with some stag's horns, and bones of the ox and horse, were found, in the year 1815, in a bed of diluvium, which is immediately incumbent on stratified beds of lias; and is composed of a mixture of various pebbles, sand, and clay : in the lower regions of which, (where the clay predominates,) the bones are found at the depth of 15 feet from the surface; they are not in the smallest degree mineralised, and have lost almost nothing of their weight or animal matter.   One of these heads, measuring in length two feet six inches, together with a small tusk, and molar tooth of an elephant, have, by the kindness of Henry Hakewill, Esq. (of architectural celebrity) been deposited in the Radcliff Library at Oxford.   The other and larger head, with a tooth and leg bone of the same animal, has been presented by Henry Warburton, Esq. to the Geological Society of London.   Of the remaining tusks of elephants, the largest is in the possession of G. Harris, Esq. of Rugby; and the other of J. Caldecot, Esq. of Lawford.   These tusks have all of them a considerable curvature outwards towards the point, like those of the one found entire in the ice of Tungusia.   By the kindness of Mr. Grimes, another enormous semi-circular tusk, from the same place, measuring seven feet in length, together with a highly valuable collection of the bones of rhinoceros, are deposited in the Oxford Museum *.

The remains of elephant, which I have mentioned as being found in North Wales, in the Vale of Clewyd, and near Dyserth, are attended with some peculiar circumstances; they are commonly said to occur in a lead mine, and so in fact they do; but it is a lead mine of an unusual kind, being conducted in a bed of diluvial gravel, that contains pebbles of lead, as the gravel beds of Cornwall, called stream works, contain pebbles and sand of tin ore.   Extensive lead mines of the same kind are worked in North America, between Lake Superior and the Missisipi.   It is the only case I know, in this country,

---

* Many of these latter have been engraved in vol. ii. part i. plate xiv. of Cuvier's Animaux Fossiles, from drawings by Miss Morland.   The largest and finest head I have ever seen of this species of rhinoceros was sent me from Siberia by J. Prescott, Esq. now resident at St. Petersburg; and I have presented it, through M. Cuvier, to the Museum of the Royal Garden at Paris, where they had no head of this animal.   This specimen is engraved in vol. ii. part i. plate xii. of Cuvier's Animaux Fossiles, and in the Philosophical Transactions for 1822, part i. plate iii.   In the British Museum there are two heads of the same species, one of which was presented to the late Sir Joseph Banks by the Emperor of Russia.

where lead is found under such circumstances in sufficient quantity to be worth working. It is locally called flat ore, from its occurring in flat or horizontal beds of gravel. Its occurrence here is explained by the position of this gravel bed at the mouth of a valley of denudation, cut in the limestone hills of Holkin, which are full of lead veins. The gravel resulting from this destruction contains fragments of lead ore, mixed up with the wreck of the rock, that formed its matrix before the excavation of the valley. Its thickness is unusually great, and several mines are worked in it; one, called Grovant Mine, gives the following section:

1. Vegetable mould, two feet.
2. Clay, mixed with some sand and rolled stones, 26 yards.
3. Gravel beds, containing rolled pieces of lead of all sizes, eight yards.

In another mine, called Talarcoch, the remains of ox and stag are found at present: and in 1815 a pair of stag's horns were discovered at 60 yards below the surface, and are now in the possession of the Earl of Plymouth at Tardebig. The section of this mine is:

1. Vegetable mould, two feet.
2. Clay, 26 yards.
3. Sand and gravel, 68 yards; containing pebbles of copper as well as of lead. Horns, teeth, and bones, are found in it, at from 40 to 70 yards from the surface, and also at the bottom of the gravel, in immediate contact with the subjacent limestone rock.

Another shaft dug one mile south of St. Asaph, at a spot between the Ebwy and the Clewyd, presented irregular alternations of clay and gravel, to the depth of 88 feet.   For the above particulars, as to these lead mines, I am indebted to the kindness of C. Stokes, Esq. and Robert Dawson, Esq. *

Of the occurrence of elephants in Scotland, we have the following evidence by Mr. Bald, in the 4th vol. of the Wernerian Transactions, p. 58, where he states, that an elephant's tusk, 39 inches long, and 13 in circumference, was found embedded in diluvial clay at Clifton Hall, between Edinburgh and Falkirk, in cutting the canal in July, 1820, at the depth of 15 or 20 feet below the present surface; it was in so high a state of preservation, that it was purchased for two pounds, and sawn asunder, by an ivory turner at Edinburgh, to be made into chess men; but the parts have been preserved by Sir Alexander Maitland Gibson.   Two other tusks, of nearly the same size, were also discovered, with several small bones lying near them, in Jan. 1817, at Kilmaurs, in Ayrshire, near the water of Carmel, at the depth of $17\frac{1}{2}$ feet from the surface, in a mass of similar diluvial clay.   Parts of these are preserved at Eglington Castle, and in the College Museum at Edinburgh †.

* I am informed by Professor Sedgewick, that being in Derbyshire in 1818, he was told that bones had been found in a lead mine on Bakewell Moor, nearly 100 feet below the surface; and that on visiting the spot, he found the miners working in a fissure filled with pebbles of limestone and sandstone, large rolled pebbles of galena, and mud: amongst these were teeth and fragments of the bones of horses. These probably had been all washed together into the open fissure at the same time, when the galena pebbles and elephants' bones were lodged together in the Vale of Clewyd, and the rhinoceros, &c. washed into the Dream's Cave near Wirksworth.

† The state of preservation of these tusks is nearly equal to that of the fossil ivory of Russia: those found in England are usually more decayed. The only one I have seen sufficiently hard to be used by the turners was found on the coast of Yorkshire, where the

With respect to Ireland, there is a description in the Philosophical Transactions for 1715, by Dr. Molineux, accompanied by engravings, of some molar teeth of elephant found at Maghery, in the county of Cavan; and the occurrence of the remains of the same large and extinct species of elk, with that found in the diluvial clay and gravel of Walton in Essex, and other parts of England, is notorious and almost universal in the marl that lies at the bottom of the Irish peat bogs.

For foreign localities of the fossil elephant, I have already referred to Cuvier's account of places in which they have been found all over Europe. Blumenbach, in his Archæologia Telluris, part i. p. 12, 1803, states, that within his knowledge more than 200 elephants, and 30 rhinoceroses, have been found in Germany.

At Seilberg, near Canstadt, on the Necker, in 1816, they discovered, in 24 hours, 21 teeth, or fragments of teeth of elephant, mixed with a great number of bones; and soon after, in continuing their diggings, fell on a group of 13 tusks and some molar teeth of elephants, heaped close upon each other, as if they had been packed artificially. These were all carefully removed, in their natural position, with the clay in which they were embedded, by order of the King, to the Cabinet at Stutgard. The largest of the tusks, though it had lost its point and root, was eight feet long, and one foot in diameter. They are in good preservation, and in general curved to the amount of three quarters of a circle, and bending outwards.

At the village of Thiede, on the plain, four miles south-west of

diluvium is very argillaceous: a portion of this tusk is now preserved in a Museum at Bridlington.

the town of Brunswick, a similar discovery was made in 1816, of a congeries of tusks, teeth, and bones, piled together in a heap of 10 feet square, and embedded in diluvial loam that covers some gypsum quarries in the new red sandstone. In this small heap, (see Plate XXIV.) Mr. Berger, of Brunswick, found 11 tusks of elephant, one of them 11 feet long, another 14 feet 8 inches long, and $12\frac{3}{4}$ inches in diameter, and both curved into a perfect semicircle; 30 molar teeth, and many large bones of elephant, some of which were five feet long, and one of them, according to Mr. Bieling, six feet eight inches. Mixed with these were the bones and teeth of rhinoceros, horse, ox, and stag; they all lay mixed confusedly together: none of them were rolled, or much broken; and the teeth for the most part separate and without the jaws: there were also some horns of stag. I have seen the hole from which they were taken: it remained entire in 1822, and no further search had been made in the loam surrounding it.   I saw also many of the specimens in the collection of Mr. Bieling at Brunswick, who has published a short description, with an engraving of them, as they lay in the quarry.   It is very difficult to account for this partial accumulation of various teeth and bones: they were most probably drifted together by eddies in the diluvian waters; but cannot have been rolled far, as they have rarely lost any thing of their projecting points and angles.

A third spot in which they occur in unusual abundance is near Florence, in the valley of the Arno, above the gorge of Incisa. From this gorge, the valley widens upwards to Arezzo, a distance of 25 miles, whilst the hills become gradually more and more contracted towards its lower extremity, and would meet, but for the existence of the gorge cut through them, at Incisa, and from which the town has evidently derived its name. This gorge forms the only outlet to the waters of the Arno, and appears to be of diluvial origin like that of the Derwent, at New Malton, and of the Weissent, near Muggendorf; and without it, the valley above must have been a lake: within the last ten years, parts of the skeletons of at least a hundred hippopotami have been discovered here, and placed in the Museum at Florence. With these are found also, in great abundance, the remains of rhinoceros and elephant, together with those of horses, oxen, several species of deer, hyæna, bear, tiger, fox, wolf, mastodon, hog, tapir, and beaver: they are from animals of all ages, and one of the elephants could not have been a week old. The analogies which this valley and its gorge present to those of the antediluvian lake, in the Vale of Pickering, and its gorge at Malton, as described in my account of Kirkdale, together with the resemblance of so many of the animals which at that time occupied these districts respectively, shows an identity of the antediluvian condition of Italy and England too striking to be overlooked; and each assists in throwing light on the state of the other, during that remote and obscure period in the history of our globe. For the detail of the above facts relating to the Vale of Arno, I am indebted to a communication from Mr. Pentland, who is now at Florence.

It is, however, a rare occurrence to find the remains of these animals collected in such great numbers on one small spot: the bones and teeth are more usually scattered about irregularly among the loam and gravel, and occur but seldom in entire skeletons, except in the frozen regions of Russia and Siberia; over these countries their dispersion also is universal. There is not, says Pallas, in all Asiatic Russia, from the Don to the extremity of the promontory of Tchutchis, a stream or river in the banks of which they do not find elephants and other animals now strangers to that climate. These are washed out by the violent floods arising from the thaw of the snows, and have attracted universally the attention of the natives, who collect annually the elephants' tusks to sell as ivory. I have already mentioned the elephants' teeth found by Kotzebue, in the iceberg, near Behring's Straits, and the extraordinary quantity of similar bones and teeth of elephants and oxen in the islands of mud and ice, at the mouth of the Lena. For a detailed account of these, and of the carcase found entire in the ice of Tungusia, and which is now preserved at Petersburg, I must refer to M. Cuvier's Animaux Fossiles, or to the translation of his 'Essay on the Theory of the Earth,' published by Mr. Jameson*. Mr. Mitchell, in his translation of this same Essay, has shown the extent to which this extinct species of elephant prevails in North America. Humboldt, also, has found it on the plains of Mexico, and in the province of Quito.

How is it possible to explain the general dispersion of all these

---

* A translation of the account given of this animal, in the Memoirs of the Imperial Academy of Sciences, at Petersburg, accompanied by an engraving of the entire skeleton, with the flesh still adhering to the head, has been published in a small Tract of 15 pages, by Mr. Phillips, George-yard, Lombard-street, 1819. Its two tusks together weighed 360 pounds.

remains, but by admitting that the elephants as well as all the other creatures whose bones are buried with them, were the antediluvian inhabitants of the extensive tracts of country over which we have been tracing them? and that they were all destroyed together, by the waters of the same inundation which produced the deposits of loam and gravel in which they are embedded.

EVIDENCE OF DILUVIAL ACTION AFFORDED BY DEPOSITS OF LOAM AND GRAVEL.

It is admitted on all hands, that the surface of the earth is strewed over with deposits of gravel, sand, and loam, which have been drifted to their present place by the action of water, since the formation of the strata over which these deposits are irregularly spread: to account for these appearances, various theories have been suggested, all of which have been defective, from their attempting to refer to one common cause two distinct classes of phenomena; viz. 1st, the general dispersion of gravel and loam over hills and elevated plains, as well as valleys; and 2d, the partial collection of gravel at the foot of torrents, and of mud at the mouths and along the course of rivers.   The former of these I shall endeavour to show are the effects of an universal and transient deluge, the latter are clearly referable to the action of existing causes.   I know not any work in which this distinction is so well and so clearly laid down as in a paper by Mr. Bald, (in the third volume of the Wernerian Memoirs, p. 123, and fourth volume, p. 58), in his account of the Clackmannan Coal Field; in which he says, " The alluvial cover which rests upon the rocks of this district is of two very distinct kinds, which are termed the old and the recent alluvial covers;" and this observation, he adds, applies to every district of Great Britain which he has examined: " that termed recent is found along banks of rivers and lakes, and is generally very fertile; and along the Firth of Forth is in some places

B B

90 feet deep[*]: it contains abundance of organic remains of trees, shells, &c., and is visibly forming every day: on the other hand, the old alluvial cover is of vast extent, occupying a large portion of the surface of Great Britain, is found at great heights and also under the level of the sea, and is of three kinds: first, sand; second, gravel; third, clay; the clay is sometimes mixed with sand, gravel, and boulder stones, which are several tons in weight. The whole is without horizontal divisions into beds or strata, and both large and small boulder stones are found mixed irregularly through every part of it. In some places it attains the thickness of 160 feet; besides boulder stones, it contains gravel (i. e. rounded fragments) of almost every kind of rocks, and angular fragments of the adjacent rocks, which are often of a softer nature than those which have been rolled to pebbles. It is this old alluvial cover in which the elephants' tusks are found; but besides these and the bones found with them, it has been observed to contain no other kind of organic remains; the absence of such remains, and irregular manner in which the materials of this deposit are mixed together, lead to a conclusion that they were collected by some violent and sudden convulsion totally different from the daily and gradual process by which the present alluvium has been and continues to be formed."

---

[*] It was in this alluvium, that the entire skeleton of a large whale, which is now in the College Museum, at Edinburgh, was found a few years since; it must have been drifted and stranded there, while this part of the estuary was under the process of filling up by the deposits of the present sea. The bones of whales have been found in a nearly similar position at Pentuan, in an ancient estuary that is now filled up on the coast of Cornwall. A description of the stream works at this place is given in Vol. IV. of the Geol. Trans. p. 404.

The difference between the two species of alluvium thus clearly pointed out by Mr. Bald, as prevailing in Scotland, is the same which I have stated to exist universally over the world; and for the purpose of distinguishing which, I have, in my table of the super-position of the strata in the British islands, proposed to limit the term alluvial to those partial deposits which Mr. Bald calls " recent alluvial covers," the origin of which may be referred to the daily action of torrents, rivers, and lakes; and to appropriate the term diluvium to those universal deposits of gravel and loam which he calls " the old alluvial covers," to the production of which no cause at present in action is adequate, and which can only be referred to the waters of a sudden and transient deluge*.

* The Hon. Wm. Strangways, in a valuable Synoptic Table of the Formations near Petersburg, published by him during his late residence in that city, has adopted the division I am now speaking of between diluvian and postdiluvian formations; distinguishing them by the name of diluvium and alluvium. He dates the commencement of the alluvium from the period of the retreat of the last waters that have covered the earth, and includes under it—1, Drift sand; marine, or inland; 2, Marsh land; composed of mud deposited by rivers; 3, Peat; 4, Calcareous tuf. All these formations are referable to causes that are still in daily action. Under the term diluvium he includes the superficial gravel beds that lie indiscriminately on all strata of antediluvian origin, and are composed of a mixed detritus of pebbles, sand, and clay, torn down from formations   all ages, except alluvial; and also the blocks of granite and other fragments of primitive and secondary rocks, that are scattered over the plains and low hills of that part of the north of Europe, either mixed with the superficial gravel, or lying insulated in situations to which they must have been drifted from very considerable distances, as there is no matrix near them from which they could by possibility have been derived. He has omitted to mention beds of gravel produced locally by torrents and rapid rivers, because the flat condition of the district on which his synoptic table is founded has allowed no gravel of this kind to be transported to so great a distance from the hills or mountains, from the daily detritus of which it is immediately derived.

A well digested and valuable comparative account of the mode of action and effect

I have seen a good example of these two deposits in Holland, in immediate contact with one another. The alluvial detritus of

of rivers and mountain torrents, showing that the maximum of their force is wholly incompetent either to excavate the main trunks of the valleys through which they flow, or to produce the gravel beds that cover them at a distance from the hills and mountains whence this gravel has been transported, is given in chap. 20 of Dr. Kidd's Geological Essays, and in the second Essay of Mr. Greenough's Examination of the first Principles of Geology. The same subject has been treated with equal accuracy and ability by M. Brongniart in the latter part of his "Histoire naturelle de l'Eau," published in the 14th volume of the "Dictionnaire des Sciences naturelles," and separately as a small pamphlet, which I strongly recommend to the attention of those persons who wish for correct information as to the effects produced by water upon the surface of our globe.

A difficulty occurs frequently along the base of a mountain chain, in marking the exact line which separates the deposits of postdiluvian detritus, which have been and still continue to be drifted down by wintry torrents, from that gravel which is strictly of diluvian origin. The bursting of an Alpine Lake (such as occurred in June, 1818, in the valley of the Dranse in Switzerland), and the daily action of torrents and rapid rivers in times of flood, are competent to produce partially over a limited district, beds of gravel somewhat similar to those of the great diluvian waters. Striking examples of this kind are afforded in the Duchy of Venice, along the base of that part of the Alps which lies immediately on the north and north-west of the town of Valvasone, where the flood waters of the Tagliamento, the Meduna, and the Zelline, have strewed the plains to an extent of many miles from the base of the mountains with a beach of pebbles of enormous breadth, which in summer is dry and barren, resembling a naked chesil bank on the sea shore. Similar features are presented by the Torre and Malina torrents on the north-east of Udina, and by the numerous torrents that descend into the plain of Lombardy, from the mountains on the north of Verona and Vicenza. The Trebbia and Taro rivers also, and all the torrents adjacent to them, which fall into the Po from the south, near Parma and Piacenza, cover the lands in the vicinity of their courses with a similar and annually increasing accumulation of detritus, from that part of the Apennines in which they take their origin. And in our own country, the small river of Avon Lwyd or Tor Vaen, which descends from the west side of the Blorenge mountain in Monmouthshire, by the valley of Pontypool, presents, at the point where it leaves the mountains immediately below that town, a naked strand of pebbles, that is perpetually shifting and laying desolate the level regions that succeed immediatley to the sudden termination of the steep and precipitous mountain valley, along which the torrent has its course above Pontypool.

At the point where transverse mountain valleys fall into the great longitudinal valleys,

modern rivers which is so enormous in that country, never rises above
the level of the highest possible land floods; but beneath this level
forms nearly the entire surface of that low and extensive flat; whilst
the diluvial deposits rise from beneath it into a chain of hills, composed of gravel, sand, and loam, which cross Guelderland, between
the Yssel and the Rhine, from the south-east border of the Zuyder
Zee, to Arnheim, and Nymegen, and form at the latter place a cliff,
overhanging the left bank of the Waal, and another cliff of the same
kind on the right bank of the Rhine, from Arnheim to Amerongen
on the road to Utrecht.   In the districts that lie below the flood-
level of these rivers, it is probable, that there is an extensive deposit
of this same diluvium buried beneath the alluvium, which forms the
surface; and the certainty of this fact has been established in several
places, where, from the bursting of dykes, the water has made excavations through the alluvium into the subjacent diluvium, and washed

we sometimes find a considerable talus-shaped accumulation of postdiluvian gravel, partially filling up the diluvian gorge of the transverse valley, and protruding itself to a considerable distance into the main trunk of the longitudinal valley; many striking examples of this
latter kind may be seen in ascending the passage of Mont Cenis on its western side from
Aiguebelle upwards. Here, at the termination or mouth of the transverse valleys that fall
into the main valley of the Arc, immense talusses of gravel of modern origin project into
the latter valley, being often incumbent on, and easily distinguishable from, the subjacent
beds of diluvial gravel, and sometimes protruding across the great longitudinal valley, so
as entirely to obstruct it, with the exception of a small passage which is kept open by the
present river, in the lowest edge of the talus.   I have seen also similar examples well
displayed at the mouths of the transverse valleys, that fall into the main valley of the
Kiszucza River, on the Hungarian side of the pass of Iablunka, at the west extremity of
the Carpathians.   Deposits of this kind go on accumulating daily under our observation,
and may, by careful investigation, be always distinguished from that gravel which is
strictly diluvian.

up from it the teeth and bones of the extinct elephant* and other animals, which are peculiar to that formation.

The term alluvial has, however, been hitherto applied too generally, not only to the diluvial and postdiluvial formations I am now speaking of, but also to all deposits of whatever era, in the regular strata, that have been drifted to their present place by the action of water; and when thus used affords no kind of information as to the age or relations of the deposit to which it is applied.

The important distinction I have been drawing between diluvium and alluvium is not less remarkable in the lowlands of the estuaries of the Thames, the Wash of Lincolnshire, and the Humber, than it is in Holland. At the mouths of all these rivers, and of others less important, there is a continual gain of new land, by depositions of mud and silt, analogous to those which form deltas at the mouth of the Rhine, the Po, and the Nile. On the east coast of England, there is also a considerable addition of silt and mud on some parts, which is derived from extensive cliffs of diluvial clay and mud, that are continually cut away by the action of the sea in others. The history of deposits of this kind has been so admirably illustrated in M. Cuvier's Theory of the Earth, and the proofs he advances to show that the period at which they began to be formed cannot have been exceedingly remote, are so decisive, that, referring my readers to him for further information on this subject, I proceed to consider the evidences of

---

* In the Museum at Leyden, there is an immense os innominatum of an elephant, three feet six inches long, which was washed up in this manner in 1809, by the inundation at Leonen, in the district of Betuwe. The head of an elephant three feet ten inches long was discovered in a similar manner after an inundation at Heukelum.

diluvial action preceding the commencement of this alluvium, viz. the history of the formation and extent of those deposits of loam and gravel which I have already to a certain degree marked out in tracing that of the elephants embedded in them, and which, as I have before stated, are of universal occurrence in this as in every other country.

The loam itself possesses no character by which it is easy to ascertain the source from which it has been derived, but usually varies with the nature of the hills composing the adjacent districts. It is of immense extent on the Continent, is known in Germany by the appellation of " Dammerde," and of " Terrain d'attrissement" in France; and its occasional abundance on the chalk of the north of France is the cause of greater fertility in some of the chalky districts of that country than of our own. But the deposits of gravel contain solid fragments, and often large blocks of granite and other rocks, which can be traced to their parent mountain ; the position of which with respect to the fragments is important, as affording a proof of the direction of the currents that drifted them to their present place of lodgement.

This diluvial gravel is almost always of a compound character, containing amongst the detritus of each immediate neighbourhood, which usually forms its greatest bulk, rolled fragments of rocks, whose native bed occurs only at great distances, and which must have been drifted thence at the time of the formation of the gravel, in which they are at present lodged.

Now, if we examine with this view the eastern coast of England, we shall find that from the mouth of the Tweed to that of the Thames, it is covered irregularly with beds of superficial loam, or clay and gravel, of enormous thickness, not only in the lowland districts, but also on the summits of lofty hills, and on the elevated table lands of the interior: e. g. on the north of Bridlington there are beds of this kind forming a cap, on the chalk hills and cliffs between that town and Flamborough Head; they are also found, in a similar position, between Flamborough Head and Filey Bay, as well as on the top of the cliffs near Scarborough, and on the north of Tynemouth. They also occupy the whole coast of the Holderness part of Yorkshire and the coast of Lincolnshire, and form the entire district between the sea and the wolds of these two counties. They abound still further on the shores of Norfolk, Suffolk, and Essex, forming along all this coast cliffs that overhang the sea, and are undergoing a perpetual destruction by the waves, so that many villages have been lost, and the ground on which they stood reduced below the level of the present sea. In the interior, they are spread widely over the table-lands of Suffolk and Norfolk; and in the north cover much of the country between Newcastle and Tynemouth, and between Stockton-on-Tees and Darlington *.

Their most common character in the localities here enumerated is that of a tough bluish clay, through which are dispersed irregularly pebbles of various kinds, together with the bones of elephants and other animals before spoken of. The pebbles are of two classes, 1. composed of the wreck of the adjacent inland districts of England; 2. large blocks and pebbles of many varieties of primitive and transition rocks which do not occur in England, and which can only be accounted for by supposing them to have been drifted from the

---

* It is well displayed in the quarries where the great Whin dyke crosses the Tees, a few miles above Stockton.

nearest continental strata of Norway, by a force of water analogous to and contemporaneous with that which drifted the blocks of Finland granite over the plains of Russia, and the North of Germany. A diluvial current from the North is the only adequate cause that can be proposed, and it is one that seems to satisfy all the conditions of our problem.

The pebbles of iridescent felspar, like that of Labrador, which are found on the coast near Bridlington, and resemble similar fragments near Petersburg, can only be referred to the primitive districts of the most northern parts of Europe. Many of the other pebbles of the English coast can be identified with rocks that are known to exist in Norway, and must have been drifted hither at the time of the deposition of the masses of clay and gravel through which they are disseminated; it is impossible to refer them to any action of the present sea, because they occur on the high table lands of the interior as well as on the coasts, and because the cliffs themselves, being composed of clay mixed with the pebbles in question, are undergoing daily destruction, and receive no addition from the action of the present waves.

These foreign and probably Norwegian pebbles on the coast of England are mixed up with the wreck of the hills composing the interior of each district respectively; and the component fragments of the latter are less rolled and more angular than those which have come from the Continent: thus in the counties of Norfolk, Suffolk, Lincoln, and Yorkshire, the diluvium contains a large proportion of fragments of chalk and chalk-flints, derivative from the immediate neighbourhood, and not much rolled; whilst in the counties of Durham and Northumberland there are no remains of chalk, but a similar admixture of the wreck of strata that compose the coal formation of these counties. In the diluvium of the numerous valleys of Yorkshire, that unite to fall into the Humber, there is a similar admixture of the debris of strata composing the adjacent country, with rounded fragments of distant rocks; and in the county of Durham I collected, within a few miles on the north of Darlington, pebbles of more than 20 varieties of slate and greenstone rocks, that occur no where nearer than the Lake district of Cumberland. In the street at Darlington, at the north end of the town, is a large block of granite, of the same variety with those at Shap, near Penrith. Blocks of the same granite lie in the bed of the Tees at Bernard Castle, and near the highest points of the pass of Stainmoor Forest. Similar blocks are found also on the elevated plain of Sedgefield, on the south-east of Durham, in all these places they are mixed with blocks of greenstone.

The nearest point from which these blocks and pebbles could possibly have been derived is the Lake district of Cumberland; and the only place in which this peculiar kind of granite occurs in situ is the neighbourhood of Shap, just mentioned, from which they are at present separated by the lofty ridge and escarpment of Cross Fell and Stainmoor Forest. If the difficulty of transporting them over this barrier be thought too great, the only remaining solution will be that they have come from Norway, like the other pebbles before mentioned, as abounding in the diluvium of the whole east coast of England. I am disposed myself to adopt the opinion of Sir James Hall, that they have come from Cumberland.

In the valley of the Trent, north-east of Newark, I have noticed a similar admixture of pebbles of primitive and transition rocks, with rounded and with angular chalk flints, that may have come from the wolds of Lincolnshire.   At Chellaston also, on the south of Derby, higher up in the same valley, I found the gypsum quarries to be buried beneath a thick bed of diluvial clay, through which are dispersed angular fragments of lias, oolite, hard chalk, and chalk flints, drifted from no great distance, and confusedly mixed with highly rolled pebbles of quartz, and other transition rocks.

Mr. Farey, in his Agricultural Report of Derbyshire, gives a long and interesting list of the deposits of gravel in that county, from which it appears, that fragments of all the English formations, from granite upwards to chalk, are accumulated abundantly in the form of diluvial gravel in that midland part of England, and I have myself found plenty of chalk flints in the gravel pits three miles north-west of the town of Derby.

It is mentioned, in a paper by Mr. Aikin, on the gravel at Litchfield, in the 4th vol. of the Geological Transactions, that he found near that town pebbles of granite, syenite, greenstone, schist, limestone, quartz, chalcedony, and hornstone : amongst these the pebbles of coralline limestone are most abundant, and like those of hornstone appear to have been derived from the mountain limestone of Derbyshire.

The Rev. W. D. Conybeare has noticed in the following terms the superficial gravel beds of the midland districts of Leicester, Rutland, and Buckinghamshire.

" The gravel accumulated in the midland counties of England," says he, " is worthy of much more attention than it has hitherto received. These accumulations extend over the plains that lie between the north-west escarpment of the great oolite chain, and also over the low tract between these hills and the north-west escarpment of the chalk of Bucks, Herts, and Bedfordshire; but they are more particularly abundant in the former position, where extending many fathoms in depth, they often effectually conceal the subjacent strata, and sometimes by their acervation constitute decided hills. Tracts of this description are particularly abundant on the borders of Rutland, Warwick, and Leicestershire. From Houghton on the Hill, near Leicester, to Braunston, near Daventry, proceeding by Market Harborough and Lutterworth, the traveller passes over a continuous bed of gravel for about 40 miles. Near Hinckley, great depositions of gravel, probably connected with this mass, are found, and afford pebbles containing specimens of the organic remains of most of the secondary strata in England. This deposition may probably be traced continuously to that of Shipston-on-Stour, most of the hillocks scattered over the lias and red marl tract, between Southam and Shipston, being crowned with this gravel.

" These accumulations of pebbles, promiscuously heaped together, are composed of the wreck of rocks of the most distant ages, and which exist in their native state only in distant quarters of the island. Flints from the chalk formation, accompanied by rounded masses of hard chalk, and fragments of the different oolite rocks, seem, however, decidedly predominant in Leicestershire; and next to these in quantity are the granular quartz rock pebbles, resembling those from the Lickey, with others of white quartz, and dark coloured hard flinty slate.   It would, however, be not difficult in many places, as for instance on the west of Market Harborough, and in the Valley of Shipston-on-Stour, to form almost a complete geological series of English rocks from among these rounded fragments, which often occur in boulders of very considerable size.

" The immense quantities of fragments of chalk flints scattered in this gravel at such a distance from the present limits of the chalk, is a very observable circumstance, and seems decisively to indicate that this formation must once have occupied a much wider space than it does at present.   Near Sywell, six miles north-east of North-ampton, on the oolite formation, are some fields as thickly strewed over with fragments of pure white chalk, as the surface of stony arable land is usually with the substance of the subjacent rock.   Even as far as Derbyshire, chalk flints are commonly found dispersed over the surface of the country."

The accumulations of gravel on the low grounds along the valley of Buckingham and Bedford are principally composed of fragments of the neighbouring rocks of oolite and chalk, with an occasional ad-mixture of quartz or pebbles from the central counties.   I have also noticed them in Whittlebury Forest near Northampton, and at Brackley.   The late Sir Joseph Banks informed me he had observed pebbles of porphyry in the road gravel on the north side of the town of Dunstable.   There is no nearer place from which these latter could have been derived than the porphyritic rocks of Charnwood Forest, in Leicestershire.

Professor Sedgwick has ascertained, that the gravel beds on the summit of the Gogmagog Chalk Hills, near Cambridge, and on the hills adjoining towards Bedfordshire, as well as that in the valleys, contain not only the wreck of chalk strata, but also fragments of almost every formation that occurs in England; amongst them he has found the joint of a basaltic pillar, between one and two feet long.

Another striking example of a similar kind is afforded by the gravel of the valleys of the Thames from London to Oxford, and of the Cherwell and Evenlode, that fall into the Thames from the northern parts of Oxfordshire. (See Plate XXVII.) I shall subjoin, in an appendix, a detailed account of this gravel, and of the state of the hills and valleys over which it is dispersed, extracted from a paper I have published on the Lickey Hill, in the 5th vol. of the Geological Transactions. Its phenomena are in perfect unison with all the other cases I have been examining, and show the effect of a violent rush of waters from the north, which has drifted pebbles of quartz rock from the plains of Warwickshire, and other central counties, over the whole country intermediate between them and London, along the line of these three rivers; and has mixed them up in each district with the angular and slightly rolled detritus of the adjacent hills, so that we have pebbles of the porphyry and greenstone of Charnwood Forest, at Abingdon, and Oxford; and pebbles of the rocks near Birmingham, at Maidenhead, and in Hyde Park.

It appears then we have evidence, that a current from the north has drifted to their present place, along the whole east coast of England, that portion of the pebbles there occurring, which cannot have been derived from this country; a certain number of them may possibly have come from the coast of Scotland, but the greater part have apparently been drifted from the other side of the German Ocean. It appears also that there are proofs of a similar current having passed over the central and south-eastern parts of England; and if we examine its western side, we find similar evidence of a violent rush of waters from the north, in the pebbles and blocks of granite and sienite of a very peculiar character, that have been drifted from the Criffle Mountain in Galloway, across Solway Frith, to the north base of the mountains of Cumberland, where I have seen them at a spot called Shalk, between Ireby and Carlisle; whilst pebbles and large blocks of another kind of granite have been drifted in still greater numbers from Ravenglass, on the west of Cumberland, over the plains of Lancashire, Cheshire, and Staffordshire: their course is marked in Mr. Greenough's map of England, and they lie in masses of some tons weight on the west of the towns of Macclesfield and Stafford, and between Dudley and Bridgnorth.

In an appendix, I shall subjoin the details of the evidence we find in the south-west of England, to show the excavation of valleys along the coast of Devon and Dorset, by the denuding force of the same diluvian waters, whose effects we have been tracing in the eastern, western, and central parts of the island.   The breadth and depth of these valleys, from Sidmouth to Bridport, may be seen, by reference to the views of their terminations in the coast at Plate XXV. whilst their origin and extent in the interior are marked in the map at Plate XXVI.   The highest summits of this district are also strewed over with rolled pebbles of quartz, that must have come from some distant part of the country, before the excavation of the valleys that now intersect it, and probably at the same time with, and by the agency of the same inundation from the north, which has drifted southwards the pebbles I have already traced over so large a portion of England.

PROOFS OF DILUVIAL ACTION IN SCOTLAND.

I will now proceed to consider the evidence we have of a similar inundation producing similar effects in Scotland.

Colonel Imrie, in his Geological Account of the Campsey Hills, published in the second volume of the Transactions of the Wernerian Society, page 35, has described a series of phenomena resulting from diluvial action, in the southern district of Stirlingshire.

He refers the removal of certain portions of the trap-rocks, which generally form the incumbent stratum of the Campsey district, to the effect of heavy and rapid currents of water, and finds many parts of their actual surface to be strewed over with an admixture of drifted clay and rolled pebbles, analogous to that which occurs along the east coast of England from Essex to Northumberland, and to bear marks of violence from the friction of heavy masses of stone that have been drifted over it.

" In all situations of this district," says he, " where the trap has disappeared, the vegetable or surface soil rests upon a strongly tenacious blue clay, much mixed with water-worn stones, and this blue clay rests upon sandstone. Among the water-worn stones imbedded in the clay, I seldom found specimens of the native rocks

D D

of the district: those which I examined consisted mostly of rocks, generally deemed of the oldest formations, such as quartz, porphyries, granites, &c.; the native beds of which are far distant to the north and west of that part of the country.

" The disappearance of the trap in some of the glens and narrow vales seems to have been produced by the attrition of heavy bodies, set in motion by a great force of water in rapid movement.

" In some of the glens and narrow vales where the trap had not entirely disappeared, I perceived upon its surface strong indications and marks of attrition.   In some places the surface of the trap was smooth, and had evidently received a considerable degree of polish; and this polish is almost always seen marked by long lineal scratches. In other places there appeared narrow grooves, apparently formed by the rapid movement of large masses of rock having swept along its surface.

" In the eastern part of the district there occurs a small elevated plain, slightly undulated.   Here the surface of the trap in some places had lost its covering of soil, and was left bare for inspection.   Upon this plain I again detected some of these scratches.   Upon the surface there were scattered immense masses of trap, which, from their apparent weight, seemed perfectly capable of forming these scratches and grooves above described, had they been put in motion and impelled along the surface.   Upon examining some of these huge masses, I found their surfaces scratched and worn in such a way as to prove sufficiently indicative to me, that they had been long subjected to attrition in water; and I also observed, that many of them presented their principal or most projecting angle towards the west, and sometimes towards the north-west; which, according to my opinion, strongly implies the direction of the current which left them in the position in which they now rest. It is not the object of this paper to dip into the causes of this phenomena; but that such currents, as were capable of the effects which I have endeavoured to describe, have overflowed the surface of our globe, is to me clearly evident; and these scratches and grooves here mentioned are some of the minor, but clear proofs of its action."

In a very able and ingenious paper by Sir James Hall, in vol. vii. of the Transactions of the Royal Society of Edinburgh, he has recorded his discovery of similar traces of the action of a mighty current on the surface of the hills and valleys near that city. These districts are not only strewed over with the gravelly wreck of rocks that have been drifted to a great distance from their native bed by the force of violent waters; but Sir James has also observed channels and furrows, which he calls dressings, still remaining on the surface of the hard rocks over which these waters passed, driving before them blocks and fragments of every substance that lay in the line of their course, and also excavating deep valleys. Where a mass of rock has been followed under ground to where its surface has been protected by a covering of clay, it is found to resemble a wet road along which a number of heavy and irregular bodies have been recently dragged; indicating that every block that passed had left its trace behind. Occasionally the scratches deviate from the general direction, but the majority agree in parallelism with each other, and with the general direction not only of the scoops and grooves of the rock upon which they occur, but also of the ridges and large features of the district. The shape of many of the valleys, moreover, is exactly similar to that of those we see excavated by water on a sand-bank in a running brook; whilst the form and relative position of the hills resemble those presented by the residuary portions of the same recent sand-banks in which the brook cuts out its little furrows.

In a small map of the immediate neighbourhood of Costorphine Hill, on the west of Edinburgh, he gives in detail fifteen examples, within a circle of two miles in diameter, each exhibiting both the large and small features which indicate the action of water flowing with violence along the surface, and carrying large blocks of stone along with it, and points out many others in the adjacent country; and after showing that it is quite impossible to refer these effects to any causes now in action, arrives at the following conclusion. " All the diluvian facts in this neighbourhood that have come under my observation concur in denoting *one inundation*, overwhelming the solid mass of this district, this inundation being the last catastrophe to which it has been exposed."

The direction of the current producing these effects in the immediate neighbourhood of Edinburgh has been from the west, and seems to have been influenced by the local circumstances of the hills enclosing the estuary of the Forth; but in the neighbourhood of Stirling, and on the east coast of Berwickshire, and generally where derivations have not arisen from local causes, it appears to have been from the north-west*.

* The whole of this paper is so very accurate and satisfactory, that I strongly recommend the perusal of it to the attention of every one who has the smallest doubts as to the evidence there is to prove that the surface of the earth owes its last form not to the gradual action of existing causes, but to the excavating force of a suddenly overwhelming and transient mass of waters.

EVIDENCE OF DILUVIAL ACTION IN WALES.

In North Wales Mr. Underwood has noticed a similar effect of diluvial scratches and scorings on the surface of the slate-rock, where it is immediately covered with a very thick bed of gravel, in a section of the new Irish road at Dynas Hill, about one mile east of Betwysycoed, near Llanrwst, and about a mile and half east of the bridge called Waterloo Bridge. It is a case completely similar to those I have quoted from Sir James Hall, and Colonel Imrie.

The deposits of gravel I have already mentioned in the Vale of Clewyd, and the valleys excavated in the hills adjacent to it, are equally of diluvial origin; and the whole low country of South Wales, from the shore of the Severn for several miles inland, is strewed over irregularly with large boulders and beds of gravel, derivative by diluvial action from the mountains that flank this district on the north. The surface and sides of these mountains are also intersected by very deep and narrow valleys of denudation; a good example of which is afforded by the valleys of the Taff, the Ebwy, and the Neath.

EVIDENCE OF DILUVIAL ACTION IN IRELAND.

With respect to Ireland, I shall adduce a few facts only, to show that it presents diluvial phenomena altogether identical with those in England and Scotland.

In the Philosophical Transactions for 1808, Dr. Richardson has pointed out the effect of mighty currents of water in excavating valleys, many hundred feet deep, in the county of Antrim, and adjoining parts of the north of Ireland; and Mr. Weaver, in his very valuable Memoir on the Geology of the East of Ireland, in the fifth volume of the Geological Transactions, p. 294, has discussed with his usual accuracy the subject of the excavation of valleys, and the gravel produced by denudation in the counties of Wicklow and Wexford.

" It seems impossible," says he, " to consider the form of any considerable portion of the surface of the earth, or to reflect even upon the nature and disposition of its alluvial tracts, without recognising the powerful agency of an agitated fluid in a state of retrocession. The abrupt and curved outlines, the fractured surfaces and denudations of extensive tracts, the sinuosities of glens, defiles, and valleys, the salient and re-entering angles, the plains, all betray its course and moulding force. To ascribe such appearances to a gradual degradation produced by the influence of the atmosphere and the current of streams, seems to be assuming causes wholly inadequate to such effects." "The flanks of the central part of the mountain chain of Wicklow and Carlow are strewed with native debris; and these are dispersed over lower ranges for a distance of some miles from the central group, and sometimes under circumstances that claim particular attention. Cronebane Hill (being composed of slate) bears upon its summit a boulder of granite (called motty-stone) $9\frac{1}{2}$ feet high, and 42 feet in circumference, and the sides of the hill are also strewed with boulders of granite nearly as large. How then did these attain their present position? The nearest granite rock is that which extends from the eastern bank of the Avonmore, towards West Aston Hill; but this is very dissimilar in aspect to the granite boulders on Cronebane. The next granite in point of distance is that of Ballincarrig, on the banks of the Avonbeg; yet in both these instances the granite rock is found in a situation several hundred feet lower than the summit of Cronebane."

"Whence are the limestone gravel and marl derived, which we find distributed along the coast of the counties of Wicklow and Wexford? The nearest visible limestone rock in the northern quarter is that which occurs at Williamstown and Booterstown, on the southern side of the bay of Dublin; and to the southward, the first rock of this description that appears is on the south coast of the town of Wexford."

After describing other phenomena of the same kind, he proceeds: "It is worthy of observation, that many of these deposits of limestone pebbles, gravel, and marl, are situated at distances from two to ten miles from the nearest part of the continuous calcareous tract, and at an elevation reaching to two, three, and four hundred feet higher than the existing surface of the limestone rock itself." " The limestone field also abounds in rolled calcareous masses, pebbles, gravel, sand, and marl, often raised into hillocks or long extended ridges, which seem to owe their form to the action of eddies and currents. There is scarcely any part of the extensive limestone tract in the centre of Ireland, that is not more or less marked by them. Sometimes these ridges appear like regular mounds, the work of art, forming a continued line of several miles in extent : that which passes by Maryborough, in the Queen's County, is a remarkable instance of this kind; and similar mounds, hillocks, and ridges, occur also in the counties of Meath, Westmeath, Kildare, Carlow, and other portions of the limestone field, in which the calcareous gravel and sand frequently exhibit a stratified disposition, the alternate layers being very distinct from each other."—Further details as to these ridges of limestone gravel may be seen in the Irish Bog Reports.

It is needless to adduce further evidence than this, to show the effects of diluvial action to be as unequivocally displayed in Ireland as in other parts of the British Islands. I do not recollect to have seen in England examples of such distinct and lofty ridges of diluvial gravel, as those in the limestone plains of Ireland, excepting in the level district of Holderness, on the east coast of Yorkshire. Here there are similar ridges, known locally by the name of barfs, and composed chiefly of rolled chalk flints, and a few primitive pebbles (apparently Norwegian). The most remarkable of these is near Bransburton, on the north-east of Beverley: it stretches across the plain like a vast chesil-bank on the sea-shore, being about 50 feet high, and 100 broad at the base, and nearly a furlong in length, and has at first sight the appearance of an artificial military mound of enormous magnitude: it bears marks of having been applied to military purposes, but is clearly of diluvial origin.

PROOFS OF DILUVIAL ACTION ON THE CONTINENT.

M. Cuvier and Brongniart, in their Geological Map of the Basin of Paris, show that the actual form and position of many of the hills in this district, especially on the side of Fontainbleau, where they stand insulated, and in rows parallel to the main direction of the valley of the Seine, can only be referred to the denuding force of a transient mass of waters. To the same waters must be referred also the pebbles of granite, and other distant primitive rocks that occur in this same neighbourhood, mixed with the wreck of the adjacent hills; and I must again refer my readers to M. Brongniart's excellent treatise on the natural history of water, in the 14 vol. of the Dictionnaire des Sciences Naturelles, for further and abundant evidence of a violent deluge having produced the actual form of the hills and valleys, and superficial deposits of loam and gravel that occur in France. In Buffon's History of the Epochs of Nature*, we find his description also of the state of the valleys in France, and of their forms, as derived from the excavating force of a mass of waters, to be in perfect harmony with those of the other authors I have been just quoting.

In Italy M. Brongniart has described the summit of the Superga, near Turin, to be covered with blocks of serpentine, and accu-

---

* Vol. XII. p. 159, Deux Pont edit. 1782.

mulations of sand and gravel (which he calls " terrains de transport anciens," as compared with the postdiluvian detritus of the floods of modern rivers), reposing on the regular strata of that steep and almost insulated mountain. These blocks can only have arrived at their present place by being drifted from some part of the distant Alpine chain that encircles the upper extremity of the valley of the Po.

De Saussure has recorded a valuable series of observations on the effects of what he calls the DEBACLE, or breaking up and transport of massive rocks and gravel, by an enormous rush of waters, in Switzerland. The most remarkable of these is the transport of blocks of granite from Mont Blanc * to the Jura Mountains, across the space which is now the Lake and Valley of Geneva. These effects appear to be only a larger example of that same diluvial action which we have been tracing in other countries; and which has operated in Switzerland on a scale proportionate to the magnitude of the Alpine masses on which it had to exert itself. For the detail of these effects, I must refer to Saussure's own descriptions, and to Sir James Hall's excellent paper in the Edinburgh Philosophical Transactions, before quoted; in which he offers some very ingenious, and not improbable conjectures, as to the manner in

---

* These blocks lie on the Jura Mountains, at an elevation of 2000 feet above the Lake of Geneva. Their size in some cases amounts, as in the Valley of Monetier, upon the Saleve Mountain, to 1200 cubic feet; and in the case of those on the Coteau de Boissy to 2250, and even to 10,296 cubic feet, which is the measure of the block called Pierre à Martin.

which these stupendous operations which have taken place in Switzerland, were brought about.

Baron Von Schlotheim, in his Nachtrage zur Petrefactenkunde, 1822, describes, in the following terms, the valleys of denudation which traverse the plains of Saxony, on the south-west of Leipsig. I remember to have been struck with them as remarkable in the vicinity of Jena, where they afford some of the most decided examples of valleys and gorges excavated entirely by diluvial action which I am acquainted with in Germany.

" The deep narrow valleys and defiles prevailing in the neighbourhood of Jena, in the valley of the Muhl, and further towards Drakendorf and Köstritz, clearly show the power with which the ancient waters raged when these channels were excavated, in which at present flow the Saale, the Elster, and the adjoining smaller streams. It is manifest that during the course of this operation large tracts of the limestone superincumbent on the gypsum, as well as of the new red sandstone, were torn and swept away."

M. Schlotheim is disposed to consider these as local occurrences, and to attribute them to the bursting, at successive periods, of the barrier of some fresh-water lakes. But this solution is inadmissible, unless we assume the existence of similar lakes at the head of every stream, and of every valley in the world; for there are none in which the effects of similar denudation are not apparent : and Mr. Weaver, in his comment on this passage, most judiciously remarks, that lakes in the present course of nature have a tendency to fill up, by a gradual accumulation on their bottoms, and not to burst their barriers; and that whatever antediluvian lakes and inland seas may have formerly existed, the gorges and defiles by which their waters were discharged can be referred to no physical cause at present in action, but were excavated by some extraneous and more mighty power than the waters of the lakes themselves; and " where," says he, " is such a power to be found. but in the agency of the diluvian waters?"

DILUVIAL ACTION IN NORTH AMERICA.

Dr. Bigsby, in his Memoir on the Geography and Geology of Lake Huron, which is about to appear in the 6th volume of the Geological Transactions, Part II., has traced in America the similar action of a violent flood of waters rushing also from the north, and drifting from thence blocks of various primitive rocks over the secondary and transition formations that compose the basin of that lake. He also notices in this same district, effects similar to the diluvial denudations in Europe, in the excavation of valleys, the separation of islands from the mainland, the formation of crags and serrated ridges of rocks, and the wearing away of the highest summits; and shows that these cannot be attributed to any causes now in action, or to any gradual subsidence of the waters of the lake, but must be referred to the great debacle of a flood advancing from the north. The same waters, he adds, have accumulated immense deposits of sand and gravel, in heaps and ridges, at various levels of the main shores and islands in the lake. These travelled fragments are foreign to the district they pervade, and are almost exclusively of the older class of rocks; granite, gneiss, mica slate, greenstone, porphyry, sienite, and amygdaloids, which occur not in the neighbourhood, but may be shown to have come from the north, and can many of them be traced to their original source in that direction. Between Lake Erie also and Lake Huron, he states the beaches and woods to be strewed with masses of gneiss, porphyries, conglomorates, and greenstones; and that similar blocks appear on the north coast of Lake Erie, which itself is for the most part composed of a series of clay-cliffs and sand-hills.

He moreover draws an accurate distinction between these diluvial driftings of the great debacle, and the small and usually angular debris of strata produced by causes now in existence: this latter remains unmoved nearly in the place in which it is formed, at the base of the parent-cliffs from which it has fallen. Thus the opposite shores of Pelletau, on Lake Huron, being of different formations, the one limestone, the other greenstone, each is lined with its own debris, and without admixture.

In the fifth volume, number 2, of Silliman's American Journal of Science, he also gives an excellent example of another species of postdiluvian operations, viz. the forming of terraces, like the parallel roads described by Dr. Macculloch, at Glenroy, in Scotland, around the edges of lakes and on the flanks of great rivers, at considerable elevations above the level of their present waters: terraces of this kind are not uncommon in North America. The case I now quote is that of the Valley of St. Etienne, near Malbay, in which Dr. Bigsby accompanies his description with a map; by which it appears that this valley has been a postdiluvian lake, which has lowered its level at successive periods, by the breaking down at distant intervals of the gorge through which the Malbay river now flows into the

St. Lawrence. The parallel terraces that encircle this dry valley show the number of successive stages by which the bursting of the gorge took place, and are exactly similar to those engraved in Dr. Macculloch's paper above quoted, in the fourth volume of the Geological Transactions*.

Sir Alexander Croke has informed me, that the summits of some of the highest hills in Nova Scotia, being composed of slate, are strewed over with large blocks of granite. The present position of these fragments can only be accounted for by supposing them to have been drifted from the nearest granitic districts by the same rush of waters that transported those described by Dr. Bigsby in the districts of Lake Huron and Lake Erie. And Dr. Meade, of

---

* I have myself observed a similar appearance of successive terraces flanking the valleys of the Rhine below Basel; of the Salza, at Golling, on the south of Saltzburg; of the Iser, at Munich; and on the sides of many other rivers that descend from the Alps. These terraces are all of postdiluvial origin, and are often formed on diluvial gravel, and indicate either the shores of lakes that have at successive periods burst their barriers, and lowered their level, or become entirely dry; or, where they occur on the sides of rivers, they are cliffs or escarpments cut in diluvial gravel by floods of extraordinary height, resulting either from unusual tempests, or from the bursting I have just mentioned of lakes in the higher districts from which these rivers are supplied.

The examples here mentioned, of the bursting of lakes, militate at first sight against the observation I have quoted from Mr. Weaver, that modern lakes have not a tendency to burst their barriers, but to fill up; still, however, he is right in his general rule, that such is the ordinary course of nature with respect to them. The bursting of modern lakes is of rare occurrence; and wherever it has happened, there is evidence of the fact, in its leaving parallel terraces of gravel on its ancient shores; and when we consider how very few the valleys are in which such terraces occur (the neighbourhood of Glen Roy, for example, being the only instance we know in Britain), it is obvious that these few cases are but rare exceptions to the general rule which Mr. Weaver has laid down.

F F

Philadelphia, in his account of the mineral waters of Ballaston and Saratoga, in the state of New York, about 200 miles north of that city, states, " That the surface of the ground, which is here composed of shale and limestone, is covered with large insulated masses of stone, commonly called boulders, consisting of large blocks of quartz, and rolled masses of other primitive rocks. These (he adds) must have been transmitted from the neighbouring mountains, as they are not attached to the rocks in situ, and have no connexion with them: they are found in every country, and only prove the action of an extensive flood of waters."

In this dispersion of blocks of granite and beds of gravel in North America, we have evidence of a debacle by the diluvian waters in the western hemisphere, analogous to that we have been examining in Europe; and the presence of the bones of elephants, and other animals which are common to the gravel of both continents, shows that the time of its formation was in each case the same.

In South America the sand and gravel in which they find the tin of Mexico, and such extensive deposits of grains of gold and precious stones, are composed of the diluvial wreck of mountains, in which, as their matrix, these minerals were once imbedded, and where they would have remained to the present hour, had they not been broken down and reduced to sand and gravel by the same diluvial waters that have in a similar manner overspread Europe with the detritus of its own mountains. I have already mentioned metalliferous examples of this detritus in the stream tin ore of Cornwall, and the lead ore that is similarly circumstanced in the Vale of

Clewyd. In the same gravel of Cornwall, and in that of Devon, Wales, and Scotland, small pebbles and grains of gold have occasionally been found; and in Ireland the gold mine that was worked a few years since in the county of Wicklow was simply a stream-work, in which the gold was dispersed in the form of small pebbles and sand, through a bed of gravel.

PROOFS OF DILUVIAL ACTION IN AFRICA AND ASIA.

The gold that occurs so largely in various parts of Africa is chiefly found, like that last spoken of, in the state of small rolled grains disseminated through diluvial sand and gravel; so also is the tin, which is so abundant in the peninsula of Malacca, and in Banca, and other islands adjacent to Sumatra, being simply diluvial or stream tin, like that of the gravel of Cornwall.

In Hindoostan, near Bombay, agates and onyx stones are collected as pebbles from diluvial gravel beds; whilst many of the plains in the interior of India contain amidst their gravel rolled pebbles of copper ore, in quantity sufficient to send large supplies of malachite to the eastern markets. The diamonds also of India, as well as of South America, and the precious stones of Ceylon, are found dispersed in the state of small sand and pebbles through diluvial gravel. In the same kind of gravel also are found the topaz pebbles of Cairn Gorm, in Scotland, and recently the white topaz of Bagshot Heath, near London.

The erroneous idea of the old mineralogists, that the sand of all rivers in the world contains gold, is true only of those which flow through countries that are strewed with the wreck of mountains, through which gold had been once disseminated, i. e. of primitive and transition rocks; and as most great rivers of the world have their origin in rocks of this kind, the very general dispersion of grains of gold along their course adds another fact to the many I have already advanced, to show the effects of diluvial action to be co-extensive with the surface of the whole earth.

PROOFS OF INUNDATION AT HIGH LEVELS.

It has been asserted by writers of high authority, and even by Cuvier, that the occurrence of the diluvian remains of the larger animals is limited to the lower regions and great valleys of the world; and an inference has been drawn, that the waters of the flood, by which they perished, did not cover the summits of the higher mountains *.

Against this hypothesis the following facts appear decisive. 1st. The blocks of granite, which have been transported from the heights of Mont Blanc to the Jura mountains, could not have been moved from their parent mountain, which is the highest in Europe, had not that mountain been below the level of the water by which they were so transported.

2d. The Alps and Carpathians, and all the other mountain regions I have ever visited in Europe, bear in the form of their component hills the same evidence of having been modified by the force of water, as do the hills of the lower regions of the earth; and in their valleys also, where there was space to afford it a lodgement, I have always found diluvial gravel of the same nature and origin with

---

* D'ailleurs l'inondation qui les a enfouis ne s'est point élevée au-dessus des grandes chaîns de montagnes, puisque les couches qu'elle a déposées et qui recouvrent les ossemens ne se trouvent que dans des plaines peu elevées.—Cuvier, vol. i. p. 202.

that of the plains below, and which can be clearly distinguished from the postdiluvian detritus of mountain torrents or rivers.

3d. With regard to the bones of animals that perished by this great inundation, although they have not yet been discovered in the high alpine gravel beds of Europe (which is but a negative fact), we have in America the bones of the mastodon at an elevation of 7800 feet above the sea, in the Camp de Géants, near Santa Fe de Bagota; and another species of the same genus in the Cordilleras, found by Humboldt, at the elevation of 7200 feet, near the volcano of Imbaburra, in the kingdom of Quito. Mr. Humboldt has also found the tooth of the fossil elephant, resembling that of the northern hemisphere, at Hue-huetoca, on the plain of Mexico; and if the animal remains of this era have not yet been discovered at such heights as these in Europe, let it be recollected that we have no elevated mountain plains like those in America; that our highest mountains are but narrow peaks, and ridges of small extent, when compared with the low country that surrounds them; and that if it were proved (which it is not) that the animals inhabited these highest points, it is more than probable that their carcases would have been drifted off, as the greater mass of their gravel has been, into the lower levels of the adjacent country.

But in central Asia the bones of horses and deer have been found at an elevation of 16,000 feet above the sea, in the Hymalaya mountains. The bones I am now speaking of are at the Royal College of Surgeons in London, and were sent last year to Sir E. Home, by Captain W. S. Webb, who procured them from the Chinese Tartars of Daba, who assured him that they were found in the north face of the snowy ridge of Kylas, in lat. 32°, at a spot which Captain Webb calculates to be not less than 16,000 feet high: they are only obtained from the masses that fall with the avalanches from the regions of perpetual snow, and are therefore said by the natives to have fallen from the clouds, and to be the bones of genii. Those I have seen are the astragalus, head of femur, and portion of humerus of a small species of horse, and some bones of deer; their medullary cavities and cancelli are lined, or entirely filled with white crystalline carbonate of lime, beautifully transparent, and the bone itself is white, and very absorbent to the tongue; their matrix is a grey calcareous sand, adhering firmly to the bones, and interspersed with small concretions of carbonate of lime.

The occurrence of these bones at such an enormous elevation in the regions of eternal snow, and consequently in a spot now unfrequented by such animals as the horse and deer, can, I think, be explained only by supposing them to be of antediluvian origin, and that the carcases of the animals were drifted to their present place, and lodged in sand, by the diluvial waters.

This appears to me the most probable solution that can be suggested; and, should it prove the true one, will add a still more decisive fact to those of the granite blocks drifted from the heights of Mont Blanc to the Jura, and the bones of diluvial animals found by Humboldt on the elevated plains of South America, to show that "all the high hills and the mountains under the whole heavens were covered," at the time when the last great physical change by an inundation of water took place, over the surface of the whole earth.

Thus far I have produced a various and, in my judgment, incontrovertible body of facts, to show that the whole earth has been subjected to a recent and universal inundation. The same opinion is maintained in Mr. Greenough's Examination of the First Principles of Geology, where he concludes his admirable summary of phenomena derivative from diluvial action, in all quarters of the earth, with the following important passages. " The universal diffusion of alluvial sand, gravel, &c. proves that at some time or other an inundation has taken place in all countries; and the presence of similar alluvial deposits, both organic and inorganic, in neighbouring or distant islands, though consisting often of substances foreign to the rocks of which the islands are respectively composed, makes it highly probable at least, that these deposits are products of the same inundation. The universal occurrence of mountains and valleys, and the symmetry which pervades their several branches and inosculations, are further proofs, not only that a deluge has swept over every part of the globe, but probably the same deluge." He also shows it to be highly probable, " that the order of things immediately preceding the deluge closely resembled the present order, and was suddenly interrupted by a general flood, which swept away the quadrupeds from the continents, tore up the solid strata, and reduced the surface to a state of ruin : but this disorder was of short duration ; the mutilated earth did not cease to be a planet ; animals and plants, similar to those which had perished, once more adorned its surface ; and nature again submitted to that regular system of laws, which has continued uninterrupted to the present day."

In the works of Catcott, Jones, and Hutchinson*, a mass of strong evidence is brought forward to show the agency of diluvial currents in excavating valleys, over large portions of the surface of this island. And M. Cuvier, in his Essay on the Theory of the Earth, expresses his conviction, that if there be any one fact thoroughly established by geological investigations, it is that of the low antiquity of the present state of the surface of the earth, and the circumstance of its having been overwhelmed at no very distant period by the waters of a transient deluge; and although Voltaire may have indulged himself in denying the possibility of such an event†, and Linnæus have overlooked its evidences‡, the discoveries of modern geology, founded on the accurate observation of natural phenomena, prove to

---

* See Catcott, on the Deluge, 1768; Jones's Physiological Disquisitions; and Hutchinson's Works, Vol. XII.

† " Y a-t-il eu un temps où le globe a été entièrement inondé ? Cela est physiquement impossible."—Voltaire, Dict. Phil. Art. Inondation.

‡ The opinion expressed by Linnæus, that he could discover in the earth's structure no proofs whatever of a deluge amidst abundant evidences of very high antiquity, was obvious to be adopted by an accurate observer, at a time when it was attempted to explain all the phenomena of stratification and organic remains, by reference to this single catastrophe; the infant state of geology at that time rendered it almost impossible to distinguish the phenomena which are strictly of diluvial origin from those which must be referred to other and more ancient causes: but the advances that have since been made in this science have established a numerous and widely varied series of facts, a certain class of which bears as unequivocal evidence to the existence of a deluge at no very distant period, as the phenomena of stratification afford on the other hand of more ancient revolutions affecting our planet during the time in which its strata were being deposited; and it has been from want of accuracy in distinguishing between these two distinct classes of facts that errors have arisen, such as those into which Linnæus fell. For an explanation of the manner in which these natural appearances may be reconciled with the Mosaic account of the creation, I must again refer to my inaugural lecture, before quoted.

G G

demonstration, that there has been an universal inundation of the earth, though they have not yet shown by what physical cause it was produced: and I cannot better conclude this part of my subject than by extracting from my inaugural lecture, before alluded to, the following summary of the facts to which, in addition to those afforded by the interior of caves and fissures, I now appeal.   They are as follows:

I. The general shape and position of hills and valleys; the former having their sides and surfaces universally modified by the action of violent waters, and presenting often the same alternation of salient and retiring angles that marks the course of a common river: and the latter, in those cases which are called valleys of denundation, being attended with such phenomena as show them to owe their existence entirely to excavation under the action of a flood of waters.

II. The almost universal confluence and successive inosculations of minor valleys with each other, and final termination of them all in some main trunk which conducts them to the sea; and the rare interruption of their courses by transverse barriers producing lakes.

III. The occurrence of detached insulated masses of horizontal strata, called outliers, at considerable distances from the beds of which they once evidently formed a continuous part, and from which they have been separated at a recent period by deep and precipitous valleys of denudation.

IV. The immense deposits of gravel that occur occasionally on the summit of hills, and almost universally in valleys over the whole world; in situations to which no torrents or rivers that are now in action could ever have drifted them.

V. The nature of this gravel, being in part composed of the wreck of the neighbouring hills, and partly of fragments and blocks that have been transported from distant regions.

VI. The nature and condition of the organic remains deposited in this gravel; many of them being identical with species that now exist, and very few having undergone the smallest process of mineralization. Their condition resembles that of common grave bones, being in so recent a state and having undergone so little decay, that if the records of history, and the circumstances that attend them, did not absolutely forbid such a supposition, we should be inclined to attribute them even to a much later period than the deluge : and certainly there is in my opinion no single fact connected with them, that should lead us to date their origin from any more ancient era.

VII. The total impossibility of referring any one of these appearances to the effect of ancient or modern rivers, or any other causes, that are now, or appear ever to have been in action, since the retreat of the diluvian waters.

VIII. The analogous occurrence of similar phenomena in almost all the regions of the world that have hitherto been scientifically investigated, presenting a series of facts that are uniformly consistent with the hypothesis of a contemporaneous and diluvial origin.

IX. The perfect harmony and consistency in the circumstances of those few changes that now go on, (*e. g.* the formation of ravines and gravel by mountain torrents; the limited depth and continual growth of peat bogs; the formation of tufa, sand-banks, and deltas; and the filling up of lakes, estuaries, and marshes,) with the hypothesis which dates the commencement of all such operations at a period not more ancient than that which our received chronologies assign to the deluge.

All these facts, whether considered collectively or separately, present such a conformity of proofs, tending to establish the universality of a recent inundation of the earth, as no difficulties or objections that have hitherto arisen are in any way sufficient to overrule.

In the full confidence that these difficulties will at length be removed by the further extension of physical observations, we may for the present rest satisfied with the argument that numberless phenomena have been already ascertained, which without the admission of an universal deluge, it seems not easy, nay, utterly impossible to explain.

POSTSCRIPT.

As I have ventured in this work to controvert the opinions expressed by M. Cuvier in his first edition, on points of high importance, in relation to the chronology of the animal remains contained in the caves, fissures, and diluvian gravel; I am much gratified that the recent publication of the fourth volume of his second edition enables me to subjoin the testimony of that illustrious naturalist to the correctness of my views on the points in question, and to add the flattering sanction of his full approbation of the description I have published of the cave at Kirkdale, and of the important inferences I have founded upon its phenomena.

At page 224, discussing the date of the osseous breccia of Gibraltar, and on the coast of the Mediterranean, which he had before considered to be more recent than the bones in the caves and diluvian gravel, M. Cuvier says, " Je reviens donc à l'idée que je n'avois osé embrasser autrefois; celle que ces dépôts des brèches osseuses ont été formés aux dépens de la population contemporaire des rhinocéros et des éléphans fossiles."

And again at page 486, " les brèches osseuses paroissent aujourd'hui sous un point de vue d'un intérêt tout nouveau; le nombre des espèces manifestement inconnues et des espèces au moins étrangéres, qu'elles recèlent, s'est beaucoup accru." "Ces espèces inconnues reculent l'âge d'une grande partie de ces brèches bien au delà de l'époque où on les croyoit formées, et portent à les regarder au moins comme contemporaines des couches qui renferment les os d'éléphant, de rhinocéros, et d'hippopotame."

This is precisely the evidence to which I appealed in my first account of Kirkdale, as that which would be decisive of the antiquity I wished to establish with respect to the bones of the osseous breccia at Gibraltar, and in similar fissures.

With respect also to the relative ages of the bones found in caverns, and in diluvium, M. Cuvier admits the same conclusion, page 486. "Les cavernes à ossemens réclament aujourd'hui la même antiquité. Parmi les nombreux carnassiers qui les remplissent, il en est un, l'hyène, qui s'est trouvé associé soit à Kirkdale, soit à Fouvent, soit près de Canstadt et d'Eichstadt, aux éléphans, aux rhinocéros, à narines cloisonnées, aux hippopotames, en un mot, aux grands pachydermes des terrains meubles; et comme la même espèce accompagne à Gailenreuth les tigres et les grands ours elle fait nécessairement remonter ces derniers animaux aussi haut qu'elle dans le temps."

And again at page 305, speaking of the same subject, he says, " Il est suffisamment prouvé que ces divers animaux ont vécus ensemble dans les mêmes pays et ont appartenu à la même époque. Ce fait important me paroît avoir été parfaitement établi par M. Buckland."

With respect to Kirkdale, he says, page 394, " Le depôt le plus abundant en os d'hyène que l'on ait jamais observé, où leur nombre va pour ainsi dire jusqu'au merveilleux, c'est la caverne de Kirkdale, dans le comté d'York, que j'ai décrite ci-dessus d'après M. Buckland."

And at page 302, " En général, il paroit qu'avant les dernières découvertes, et surtout celle qui vient d'être faite dans le comté d'York, on ne connoissoit guère que celles d'Allemagne, et de Hongrie, qui fussent riches en ossemens de carnassiers.

" A la vérité, on pouvoit déjà croire que le rocher de Fouvent, dont nous avons parlé dans notre premier volume, et qui montre dans une de ses cavités des os d'hyènes en même temps que d'éléphans, de rhinocéros, et de chevaux, appartenoit à cet ordre de phénomènes, mais, comme on ne pénétra point dans la profondeur, on ne put constater ce qui en étoit.

" Il n'en a pas été de même de la caverne de Kirkdale.   Visitée aussitôt après sa découverte par plusieurs hommes instruits, et surtout par le savant et ingénieux géologiste, M. Buckland, on n'a rien à desirer à son sujet."

M. Cuvier also expresses an opinion which coincides entirely with my own, " that the human race had not established themselves in those countries where the animal remains under consideration have hitherto been found, in the period preceding the grand inundation by which they were destroyed."

APPENDIX.

APPENDIX.

ON THE EXCAVATION OF VALLEYS BY DILUVIAL DENUDATION.

I HAVE reserved for this place, in the form of an Appendix, the following details respecting two districts which I have already quoted in my specification of the proofs of diluvial action in the south of England, because the particulars herein enumerated would have interrupted the course of my former argument; and also, because they have already been published in the first volume of the New Series of the Geol. Trans. Part I., and in the fifth volume of the Old Series of the same Transactions, Part II. They relate to the valleys of denudation that intersect the coast of Devon and Dorset, and to the excavation of valleys and dispersion of beds of gravel in the county of Warwick, and along the course of the Cherwell, Evenlode, and Thames, from Warwickshire to London.

H H 2

We have few opportunities of witnessing by direct experiment or observation the force of immense masses of water, in excavating hollows on the earth's surface, and removing to a great distance the fragments which they tear away ; and were it not for the ravages we occasionally see produced by such comparatively trifling causes, as the bursting of a dyke in Holland, or of the barrier of an Alpine lake, we could scarcely believe that there are valleys of many miles in breadth, and many hundred feet in depth, which owe their origin exclusively to the excavating power of a flood of waters.

Our present rivers excavate but little, as they flow through valleys already formed by an overwhelming ocean ; and the destructive action of the present sea is limited to the partial cutting away of cliffs by the slow undermining of the waves in storms and at high tides.   Yet we know from the effect of a mountain torrent in cutting ravines and drifting gravel ; from the blocks of granite which were lifted to an elevated point on the side of a mountain by the bursting of a small lake in the Val de Bagnes, in Switzerland ; and from the excavation of the Zuyderzee, by the bursting of a dyke in Holland ; that the force of water in rapid motion is competent both to transport such masses of gravel and granite blocks as we have been tracing over the world, and to excavate valleys which though many miles in breadth, and many hundred, and in some cases perhaps, some thousand feet in depth, still bear a due proportion to the bulk and power of the agent that produced them.

" When we call to mind," says Mr. Sumner, in his inestimable and most judicious work on the Records of Creation, Vol. II. p. 350, " the destruction which is spread by a sudden alteration in the level of a very inconsiderable body of water, even to the extent of 50 or 100 feet, we cannot easily assign limits to the effect of a body of waters like the ocean pouring in over the land when its level was destroyed; we are at a loss to conceive what the power of such a machine might be when once in operation."

An agent thus gigantic appears to have operated universally on the surface of our planet, at the period of the deluge; the spaces then laid bare by the sweeping away of the solid materials that had before filled them, are called valleys of denudation; and the effects we see produced by water in the minor cases I have just mentioned, by presenting us an example within tangible limits, prepare us to comprehend the mighty and stupendous magnitude of those forces, by which whole strata were swept away, and valleys laid open, and gorges excavated in the more solid portions of the substance of the earth, bearing the same proportion to the overwhelming ocean by which they were produced, that modern ravines on the sides of mountains bear to the torrents which since the retreat of the deluge have created and continue to enlarge them.

When a gorge or valley takes its beginning, and continues its whole extent within the area of strata that are horizontal, or nearly so, and which bear no mark of having been moved from their original place by elevation, depression, or disturbance of any kind; and when it is also inclosed by hills that afford an exact correspondence of opposite parts, its origin must be referred to the removal of the substances that once filled it: and as it is quite impossible that this removal could have been produced in any conceivable duration of years by rivers that now flow through them, (since all the component streams, and consequently the rivers themselves, which are made up of their aggregate, owe their existence to the prior existence of the valleys through which they flow,) we must attribute it to some cause more powerful than any at present in action, and the only admissible explanation that suggests itself is, that they were excavated by the denuding force of a transient deluge.

That these excavations took place at a period subsequent to that at which the earth was inhabited by the hyænas, bears, elephants, rhinoceroses, &c. whose remains we find in caves, and diluvial loam and gravel, is evident both from the fact that the outscourings of these valleys form the gravel in which such bones are for the most part embedded; and from the number of caves (once inhabited as dens) that have been intersected and laid open in the cliffs that flank their sides and narrow gorges. The present entrance of these caves is often a hole in an absolutely vertical precipice, which it is impossible to approach except by ropes or ladders, and which, therefore, could not have been accessible to the animals whose bones we find within, if the caves had originally terminated, as they do at present, in the face of a precipice; it follows therefore, that the creation of such precipices, and consequently of the valleys in question, was posterior to the time in which the beasts occupied these dens. See an illustration of this hypothesis in the three caves intersected by the gorge of the Esbach, at Plate XVIII: see also Plate XV., XVI., and XIX.

VALLEYS OF DENUDATION, AND DILUVIAL PEBBLES IN DORSET, DEVON, WILTS, AND BERKS.

Some of the best examples I am acquainted with of valleys thus produced exclusively by diluvial denudation occur in those parts of the coasts of Dorset and Devon which lie on the east of Lyme, and on the east of Sidmouth; and the annexed views and map will illustrate, better than any description, the point I am endeavouring to establish. In passing along this coast (see the Map and Views, Plates XXV. and XXVI.) we cross, nearly at right angles, a continual succession of hills and valleys, the southern extremities of which are abruptly terminated by the sea; the valleys gradually sloping into it, and the hills being abruptly truncated, and often overhanging the beach or undercliff, with a perpendicular precipice. The main direction of the greater number of these valleys is from north to south; that is, nearly in the direction of the dip of the strata in which they are excavated: the streams and rivers that flow through them are short and inconsiderable, and incompetent, even when flooded, to move any thing more weighty than mud and sand.

The greater number of these valleys, and of the hills that bound them, are within the limits of the north and north-west escarpment of the green sand formation; and in their continuation southward they cut down into the oolite, lias, or red marl, according as this or that formation constitutes the substratum over which the green sand

originally extended.   There is usually an exact correspondence in
the structure of the hills inclosing each valley; so that, whatever
stratum is found on one side, the same is discoverable on the other
side upon the prolongation of its plane: whenever there is a want of
correspondence in the strata on the opposite sides of a valley, it is
referable to a change in the substrata upon which the excavating
waters had to exert their force.

The section of the hills in this district usually presents an
insulated cap of chalk, or a bed of angular and unrolled chalk-flints,
reposing on a broader bed of green sand; and this again resting on
a still broader base of oolite, lias, or red marl (see Plates XXV. and
XXVI.) With the exception of the very local depression of the
chalk, and subjacent green sand, and red marl on the west of the
Axe, at Beer Cliffs, the position of the strata is regular and very
slightly inclined; nor have any subterraneous disturbances operated
to an important degree to affect the form of the valleys.

If we examine the valleys that fall into the bay of Charmouth
from Burton on the east to Exmouth on the west, viz. that of the
Bredy, the Brit, the Char, the Axe, the Sid, and the Otter, we shall
find them all to be valleys of diluvian excavation; their flanks are
similarly constructed of parallel and respectively identical beds; and
the commencements of them all originate within the area and on the
south side of the escarpment of the green sand.

The valley of the Sid, as it is coloured in the annexed map, may
from its shortness and simplicity be taken as an example of the rest;
it originates in the green sand, but soon cuts down to the red marl,
and continues upon it to the sea; in both these respects it agrees

with the upper branches of the Otter, and with the valleys that fall from the west into that of the Axe.

But in those cases where the lias and oolite formations are interposed between the red marl and green sand, the base of the valley varies with the variation of substratum; this may be seen by comparing the opposite sides of the lower valleys of the Otter, the Axe, and the Char, with the variations of their substrata, as expressed in the map.

The valley of Lyme is of equal simplicity with that of Sidmouth, and differs only in that its lower strata are composed of lias instead of red marl: but the valleys of Chideock, Bridport, and Burton, being within the area of the oolite formation, have their lower slopes composed of oolite subjacent to the green sand; whilst that of Charmouth is of a mixed nature, having its western branches in green sand reposing on lias, and in some of its eastern ramifications intersecting also the oolite. In the same manner, the valley of the Axe has lias interposed between the green sand and red marl on its east flank, but none at all on its western side, below the town of Axminster. These apparent anomalies form no exception to the general principle, that the variation of the sides of the valleys is always consistent with that which is simply referable to the variation of the substrata, on which the denuding waters had to exert their force. It is moreover such as can be explained on no other theory than that of the strata having at one time been connected continuously, across the now void spaces which constitute the valleys.

The following section, taken from a series of lias quarries on the two opposite sides of the valley of the Axe, near Axminster, will show the degree of minuteness to which this correspondence extends*:

			Ft.	In.
1. White lias...	Slaty and fissile, is used for flooring when split into slabs from two to three inches thick		2	0
Clay.				
2. Burrs.........	Rough blue building stone . . .		0	10
Clay.				
3. Cockles.......	Flat and broad blue stone, containing shells and divided into two beds, each three inches thick, with a parting of clay; is used for building.—Total.		0	10
Clay.				
4. Anvils	Blue building stone, forming a bed of irregular anvil-shaped blocks .		1	0
Clay.				
5. Graze Burrs.	Good blue building stone		0	10
Clay.				
6. Fire stone....	White building stone, used also for forming the arch-work of lime pits: it divides into two beds, each four inches thick, with a parting of clay.—Total		1	0

* The details of these quarries are particularly well known to me, as they are in the immediate neighbourhood of my native place, Axminster: when a child, I often visited them, and collected specimens of fossil shells, which first excited my attention to the subject of geology.

Clay.

7. Half-foot bed. Strong blue flagstone, the best for   Ft.   In.

       paving   0   6

Clay.

8. Foot stone .. Blue paving and building stone . .   0   10

Clay.

9. Red-size....... White lias, inclining to grey, splitting

       into two or three thin slabs, and

       used for paving and building . .   0   6

Clay.

10. Under bed .. Blue building stone, used for paving,

       and the best bed of all for steps .   0   8

Clay, varying from one to six feet.

11. White rock.. White lias, rough and rubbly through-

       out ;—not good for paving or build-

       ing, but used largely to make lime,

       which is better than that of the

       other beds for plastering and in-

       door work : the thickness of this

       bed is variable; its average is . . 30   0

All the above strata are separated by thin beds of clay, varying from four inches to a foot, and exceeding the latter thickness in one case only, viz. between Nos. 10 and 11 : but the presence and relative position of each individual stratum of stone is constant; and the specific character and uses of each bed are of practical notoriety among the masons through the district round Axminster, in which there are many and distant quarries, to any one of which the above section is equally applicable; *e. g.* to the quarries of Fox Hills on the south-east, of Waycroft on the north, and of Sisterwood, Battleford, Long Leigh, Small-ridge, Green-down, and Cox-wood, on the north-west of Axminster. There can be little doubt, therefore, that the component strata of all these quarries were originally connected in one continuous plane across the now void space which forms the valley of the Axe.

The fact of excavation is evident from simple inspection of the manner in which the valleys intersect the coast, on the east of Sidmouth and the east of Lyme, as represented in the annexed views (Plate XXIII.); and it requires but little effort, either of the eye or the imagination, to restore and fill up the lost portions of the strata, that form the flanks of the valleys of Salcomb, Dunscomb, and Branscomb, on the east of Sidmouth; or of Charmouth, Seatown, and Bridport, on the east of Lyme. By prolonging the corresponding extremities of the strata on the opposite flanks, we should entirely fill up the valleys, and only restore them to the state of continuity in which they were originally deposited.

An examination of the present extent and state of the remaining portions of the chalk formation within the district we are considering, will show to what degree the diluvian waters have probably interrupted its original continuity. The insulated mass of chalk, which at Beer Head composes the entire thickness of the cliff, rises gradually westward with a continual diminution and removal of its upper surface; till after becoming successively more and more thin on the cliffs of Branscomb, Littlecomb, and Dunscomb, it finds in the latter its present extreme western boundary: beyond this boundary, on the top of Salcomb Hill, and of all the highest table-lands and insulated summits of the interior, from the ridges that encircle the vales of Sidmouth and Honiton, to the highest summits of Blackdown, and even of the distant and insulated ridge of Haldon, on the west of the valley of Exe, beds of angular and unrolled chalk-flints (which can be identified by the numerous and characteristic organic remains which they contain) are of frequent occurrence; similar beds are found also on the green sand summits that encircle the valleys of Charmouth and Axminster; large and insulated masses of chalk also occur along the coast, from Lyme nearly to Axmouth, and in the interior at Widworthy, Membury, White Stanton, and Chard; and these at distances varying from 10 to 30 miles from the present termination of the chalk formation in Dorsetshire, though within the limits of the original escarpment of the green sand.

These facts concur to show, that there was a time when the chalk covered all those spaces on which the angular chalk-flints are at this time found; and that it probably formed a continuous, or nearly continuous, stratum, from its present termination in Dorsetshire, to Haldon, on the west of Exeter *.

* There is also reason to think that the plastic clay formation was nearly coextensive with the chalk, for on the central summits of Blackdown there are rounded pebbles of chalk-flint, which resemble those found in the gravel-beds of the plastic clay formation at Blackheath: and on the hills that encircle Sidmouth there are large blocks of a siliceous breccia, composed of chalk-flints united by a strong siliceous cement, and differing from the Hertfordshire pudding-stone only in the circumstance of the imbedded flints being mostly angular, instead of rounded, as in the stone of Hertfordshire: a variation which occurs in similar blocks of the same formation at Portisham, near Abbotsbury,

From the correspondence observed by Mr. Wm. Phillips, between the strata of Dover and the hills west of Calais*, and by Mr. De la Beche, between the strata of the coast of Dorset and Devon, and those of Normandy†, it may be inferred (after making due allowance for the possible influence of those earlier causes, which in many instances have occasioned valleys) that the English Channel is a submarine valley, which owes its origin in a great measure to diluvian excavation, the opposite sides having as much correspondence as those of any valleys on the land. According to Bouache, the depth of the Straits of Dover is on an average less than 180 feet; and from thence westward to the chops of the Channel the water gradually deepens to only 420 feet, a depth less than that of the majority of inland valleys which terminate in the Bay of Charmouth; and as ordinary valleys usually increase in depth from the sides towards their centre, so also the submarine valley of the Channel is deepest in the middle, and becomes more shallow towards either shore.

It seems probable, that a large portion of the matter dislodged from the valleys of which we have been speaking, by the diluvian waters to which they owe their origin has been drifted into the principal valley, the bed of the sea; and being subsequently carried eastward, by the superior force of the flowing above that of the ebbing tide, has formed that vast bed of pebbles known by the name of the

and elsewhere.—The argument, however, arising from the presence of these blocks and pebbles is imperfect; as it is possible, though not probable, they may have been drifted to their actual place by the diluvian waters, before the excavation of the valleys.

    * See Geol. Trans. vol. v. pp. 47, &c.   † Ibid. vol. i. of New Series, p. 89.

Chesil Bank: the principal ingredients of which are such as on the above hypothesis they might be expected to be, viz. rolled chalk-flints, and pebbles of chert; the softer parts of the materials that filled these valleys, such as chalk, sand, clay, and marl, having been floated off, and drifted far into the ocean, by the violence of the diluvian waters.

The quantity of diluvian gravel which remains lodged upon the slopes, and in the lower regions of the valleys that intersect this coast, is very considerable; but it is not probable that many animal remains will be discovered in it, because the large proportion of clay with which it usually is mixed, renders it less fit for roads than the shattered chert strata of the adjacent hills, and consequently gravel pits are seldom worked in the diluvium. Enough, however, has been done to identify its animal remains with those of the diluvian gravel of other parts of England, by the discovery of several large tusks of elephants, and teeth of rhinoceros, in the valleys of Lyme and Charmouth.

On the highest parts of Blackdown, and on the insulated summits which surround the Vale of Charmouth, I have found abundantly pebbles of fat quartz, which must have been drifted thither from some distant primitive or transition country, and carried to their actual place, before the present valleys were excavated, and the steep escarpments formed, by which these high table-lands are now on every side surrounded. These cases are precisely of the same nature with those of the blocks of granite that lie on the mountains of the Jura, and on the plains of the north of Germany and Russia, and with that of the quartzose pebbles found on the tops of the hills round Oxford and Henley; which latter I shall immediately proceed to show were drifted thither from the central parts of England, before the excavation of the present valley of the Thames*.

* In the interior of Dorset, and in the counties of Wilts and Berks, the surface of the chalk is intersected in the same manner as that part of the coast we have been examining, by deep combs and valleys of denudation. It is also, occasionally, strewed over with enormous blocks of sandstone, the wreck of strata, whose softer materials have been entirely washed away. These blocks have been long noticed by the name of Sarsden Stones, and Grey Wethers, on the downs of Wilts and Berks; and are particularly abundant near Marlborough, at Kennet on the west, and in Savernake Forest on the S. E. of that town: near the former place they cover a valley more than half a mile in length, as thickly as sheep grazing in a flock, (hence their name of Grey Wethers), and have been employed in the druidical temple of Abury, at the head of this valley; whilst Savernake Forest has probably supplied the gigantic masses used to form the pillars of the larger circles at Stonehenge. They are also found scattered in great abundance over the chalk valleys at Ashdown Park, on the S. W. of Wantage. Their present position can only be referred to the same diluvial action which removed the softer portions of the sandy strata of which these blocks originally formed a part, and which excavated the valleys, over whose bottom, as well as on the sides and summits of the adjacent hills, they are now dispersed.

VALLEYS OF DENUDATION, AND BEDS OF DILUVIAL GRAVEL, IN WARWICKSHIRE, OXON, AND MIDDLESEX.

The details of the second case I proposed to give in this appendix are those which relate to the excavation of valleys, and the dispersion of beds of gravel, in the county of Warwick, and along the course of the Cherwell, Evenlode, and Thames, from Warwickshire to Oxford and London,

The new red sand-stone formation, in the central parts of England, contains an enormous deposit of pebbles of compact granular quartz, forming large beds, which may be seen near Bridgnorth, Lichfield, and Birmingham.   They here constitute a regular stratum, subordinate to the red sand-stone, and were reduced to the state of pebbles by the action of violent waters, at or before the time of the deposition of this formation.   From this lodgment, in one of the deep and regular strata of the country, enormous quantities of the pebbles in question have been swept away by the diluvial waters, and dispersed superficially over the adjacent districts and midland counties, without any reference to the nature of the rocks that lie beneath, and mixed with fragments of other rocks, both older and younger than the red sand-stone *.

* For a detailed account of the geological history of these pebbles, and of the source from which they were originally derived, I must refer to my paper on the quartz

K K

They have also been collected in prodigious numbers along the plains subjacent to the escarpment of the oolitic limestone that crosses Warwickshire, near Shipston-on-Stour; particularly on the south of that town, at the base of Long Compton Hill. (See Plate XXVII.) They are here accompanied by pebbles of white quartz, lydian stone, gneiss, porphyry, compact felspar, trap, sand-stone of several kinds, lias, chalk, and chalk-flints.

Between Shipston and Moreton in the Marsh, they have been drifted into a kind of bay, formed by the horn-shaped headland of the Campden Hills, which projects like a pier-head some miles beyond the ordinary line of the great limestone chain of the Cotswold Hills. The mouth of this bay opens directly to the north-east, from which quarter it is probable the current which brought the pebbles in question had its direction; for on the south-east of Shipston there are pebbles of a hard red species of chalk, which occurs not unfrequently in the Wolds of Yorkshire and Lincolnshire, but is never met with in the chalk of the south or south-east of England. The nearest possible point, therefore, to which these pebbles of red chalk can be referred, is the neighbourhood of Spilsby, in Lincolnshire, whence a diluvial current flowing from the north-east would find an unobstructed passage across the plains of Leicestershire to the Bay of Shipston, and Moreton in the Marsh. With these pebbles of red chalk are others of hard and compact white chalk, such as accompanies the red chalk in the two last mentioned countries, and occurs also at Ridlington, in Rutlandshire.

rock of the Lickey Hill, in the 5th vol. of the Geological Transactions, from which this extract relating to the diluvial part of their history is transcribed.

The diluvian current thus impelled into the Bay of Shipston, from the north-east, appears to have continued its course onwards beyond the head of this bay, near Moreton in the Marsh, (see Plate XXVII.) bursting in over the lowest point of depression of the great escarpment of the limestone; and being deflected thence south-east-wards by the elevated ridge of Stow in the Wold, to have gone for-ward along the line of the vale of the Evenlode by Charlbury, till it joined that of the Thames at Ensham, five miles north-west of Oxford.

This hypothesis affords the most satisfactory explanation of the origin of the great deposits of granular quartzose pebbles, which not only cover irregularly the lower regions of the valley of the Even-lode, but are scattered abundantly over the surface of the oolite strata, where they rise to a considerable height, and form table-lands on both sides that valley along its whole extent. It also accounts for the accumulation of beds of similar pebbles on the west and south of Oxford, upon the insulated and almost conical summit of Wytham Hill, and the ridge of Bagley Wood, by their position exactly op-posite the mouth of the vale of the Evenlode, at its confluence with that of the Thames, at the very point on which the driftings eva-cuated from the former valley would be collected *. Being thus introduced within the escarpment of the oolite, and having passed along the line of the Evenlode into the country round Oxford, these quartzose pebbles have been forced onwards, and mixed up

---

* Near this same point, pebbles of clear rock crystals occur scattered over the sur-face at Ensham Heath, and are applied to the purposes of jewellery, like the Bagshot Heath diamonds, as they are commonly called, being merely small pebbles of cry-stallized quartz.

with the gravelly wreck of the neighbouring hills, in each successive district along the line of the Thames, from the vale of Oxford downwards to the gravel-beds of London, their quantity decreasing with the distance from their source; so that in Hyde Park, and the Kensington gravel pits, they are less abundant than at Oxford. I have seen them on the summit of the chalk hills round Henley, Maidenhead, High Wycomb, and Beaconfield; at Dropmore also, and in the gravel pits at Burnham; in all these last-named places the great mass of the gravel is composed of imperfectly rolled flints derived from the neighbouring chalk. They are found also mixed with chalk-flints, and slightly rounded oolitic gravel, in the valley of the Cherwell, and the plains adjacent to it, from its source at Claydon and Cherwelton, to Banbury and Oxford, *e. g.* at Steeple Aston, Heyford, Rowsham, Kirtlington, and Kidlington. At Abingdon, they occur not only in the gravel-beds of the valley, but are scattered loosely over the plains composed of various strata around that town, as well as on the hills round Newnham, Dorchester, and Wallingford. Among these pebbles, especially at Abingdon and in Bagley Wood, there are many of porphyritic green-stone and green-stone slate, which cannot have come from any nearer source than Charnwood Forest, in Leicestershire.

The occurrence of quartzose pebbles in such high situations as the top of Henley Hill and Cumnor Hill, and again on the highest summits of Witchwood Forest, and generally on the elevated plains that flank the valleys of the Evenlode, the Cherwell, and the Thames, (see Plate XXVII.) goes far to prove the recent origin of the valleys through which these rivers now flow; and compels us to refer their excavation to the denuding agency of the same diluvial waters which imported the pebbles. It seems probable that the first rush of these waters drifted in the pebbles within the great escarpment of the oolite, and strewed them over the then nearly continuous plains; and that the valleys were subsequently scooped and furrowed out by the retiring action of these same waters; for it is not easy to imagine any explanation of the fact of the pebbles being heaped together on the tops of the insulated, steep, and nearly conical hill of Wytham, and of the elevated ridge of Bagley Wood near Oxford, or on the highest crest of the oolite ridge of Witchwood Forest, and the chalky summits near Henley, unless we suppose the transport of the pebbles to those summits to have been anterior to the excavation of the valleys that now intersect and surround them. Nor is this hypothesis unsupported by the fact, that it is on the elevated plains that flank the vales of the Evenlode and Cherwell, no less than in the lower regions which form their present water-courses, that the quartzose pebbles are scattered in an almost uninterrupted line, marking distinctly the course by which they have been propelled from Warwickshire into the valley of the Thames.   (See map, Plate XXVII.)

There is another strong fact tending to prove the excavation of the valleys of the Evenlode and Cherwell, and of the Thames (in part) near Oxford, to have been subsequent to the transport of the Warwickshire pebbles, namely, the absence of pebbles of oolite in the beds of gravel just mentioned as crowning the summits of Wytham Hill and Bagley Wood. Hence we may infer that the destruction of the oolite strata was not so much the effect of the advancing deluge as of its retiring waters, cutting out valleys in the table-lands, and sides of the higher ridges, and covering them with gravel, composed partly of the wreck of the strata immediately inclosing them, and partly of pebbles, which their first rush had transported from more distant regions; and thus it will appear that the lower trunks of the valleys of the Thames, Cherwell, and Evenlode, (i. e. those portions of them which may be fairly attributed to the exclusive action of denudation, and which lie below the average level of the table-lands which flank their course,) did not exist at the time of the first advance of the waters, which brought in the pebbles from Warwickshire, but were excavated by the denuding agency which they exerted during the period of their retreat *.

If we examine the geological structure of that large portion of England which lies south-east of the escarpment of the oolite formation, along its whole extent, from the coast of Dorset to that of Yorkshire, we shall find in it no one stratum that has the smallest

---

* The excavations produced by the waters entering the low point of the oolite escarpment near Moreton have been so great, (see map, Plate XXVII.) that the head springs of the Evenlode, taking their rise from the lias strata in the vale of Moreton beyond the termination of the oolite, flow south-eastward toward Oxford (through the same gap by which the diluvian current drifted in the siliceous pebbles,) instead of falling by the much shorter course of the Stour into the valley of the Severn: and it is of importance to observe, that the Evenlode and Cherwell are the only rivers of all those which flow into the Thames, which have not their head-springs within the escarpment of the great oolite. The sources of the Cherwell, and a few of its earliest tributary streams, being similarly circumstanced to those of the Evenlode, owe their existence to similar denudations cut through the oolite strata into the clay beds of the subjacent lias, even as far south as the town of Banbury. The lowness of the oolite escarpment at the lip or gap above Banbury appears still further from its having been selected as the line by which the Oxford Canal is conducted out into the sandstone plains of Warwickshire.

resemblance to the quartzose pebbles which are accumulated near Oxford, and over the other districts which I have been describing. So that if they were not transported hither across the depression at the head of the valleys of the Evenlode and of the Cherwell, they must have passed over some more elevated point of the oolite escarpment, and have come from some still more distant part of the red sandstone plains that cross the centre of the island; and the difficulty of accounting for their origin will thus only be increased by refusing to admit the solution I have proposed.

The quartzose pebbles, which I have been tracing without interruption from Birmingham to London, had, as I have before mentioned, received their roundness before they were embedded in the red sandstone formation; their form cannot therefore be referred to friction during their short transport by the diluvian waters: indeed instances are rare where fragments even of soft rocks, which have undergone no further attrition than that of these waters, have received such an extreme degree of roundness as is found in the hard quartzose pebbles we are considering. Dr. Kidd, in his excellent description of the gravel-beds round Oxford, has well and accurately marked the differences between the completely rounded pebbles and angular fragments of which this gravel is composed; the comparatively soft fragments of the neighbouring hills being angular and but slightly rolled, and those only being completely rounded which are hard, and can be traced up to ancient gravel-beds, forming part of the red sandstone strata of Warwickshire*.

* See the preface to Kidd's Elements of Mineralogy, and chap. 17 of Kidd's Geological Essays.

Similar varieties of gravel, the one angular, the other completely rolled, and derivative from the pebble beds of the plastic clay formation, occur in the valley of the Thames near London. These rounded pebbles, like those from Warwickshire, had apparently received their attrition from the long continued action of violently agitated waters, during more early revolutions that have affected our planet; whilst the imperfectly rolled fragments are referable to the diluvian waters, which drifted them only from the neighbouring hills to their present place; and from the angular state of this and similar beds of diluvial gravel, we may infer that the inundation which produced them was of short duration.

On the south-west side of the Evenlode, the valleys that intersect the Cotswold Hills in Gloucestershire are the effect of deep denudations produced on the oolite limestone, by a volume of waters rushing over strata composed of uniform and moderately yielding materials. Any irregular projections that might have existed on the original surface would cause the waters to descend with accelerated velocity over the intermediate depressions, and to excavate that series of sweeping combs and valleys that wind with the regular flexures of a meandrous river, and present masses of land alternately advancing and retiring with all the uniformity of the salient and re-entering angles that mark the course of running water.

Striking examples of such valleys extending upwards far above the highest springs that take their rise in them, and forming vast diluvian furrows along the back of the inclined planes of the great oolite formation, may be seen in passing along the line of the Roman Fossway, from Bath to Stow in the Wold: this line, being parallel to that of the great escarpment of the Cotswold Hills, crosses nearly at right angles all the valleys that descend from them towards the south-east, into the main trunks of the Thames or Avon; and in no part of it are the features of diluvian action more strongly displayed than between North Leach and Stow in the Wold.   It is obvious that such valleys cannot possibly be attributed to the action of springs or rivers that now flow through them, since they often take their origin many miles above even the highest springs: their magnitude and depth bespeak the agency of a mass of waters infinitely more powerful than even the most violent water-spouts of modern times could produce: their form also differs entirely from the deep and precipitous ravines which are excavated by mountain torrents; and if it should be contended that the bursting of a series of water-spouts would be competent to set in action such masses of water as might have been sufficient for this effect, unless we can suppose these to have fallen universally and contemporaneously, not only over the district under consideration but over the whole earth, they will afford no solution of the phenomena of these and similar contemporaneous systems of valleys which occur on strata that are similarly circumstanced in every part of the known world.

The chalk downs of England, and the upper portions of the chalky and oolitic plains of France, are universally covered with a series of dry valleys exactly similar to those that occur on the back of the inclined planes of oolite of the Cotswold Hills; and the uniform texture and moderate degree of inclination which usually attends both these formations will explain the regularity of the diluvian valleys that have been excavated on their surface.

In strata of higher antiquity, that have been more shattered and disturbed by violent convulsions (i. e. in the coal formation, and also in transition and primitive rocks), irregularities of texture and disposition in the strata on which the diluvian waters had to exert their force have caused the features of the valleys that traverse them to be much less exclusively derivative from the simple action of a retiring flood of waters; and indeed have rendered the form, inclination, hardness, and relative position of the masses on which these waters had to operate, essential elements of any accurate calculations as to the quantity of effect that must be referred to them. Though traces of diluvial action are most unequivocally visible over the surface of the whole earth, we must not attribute the origin of all valleys exclusively to that action; in such cases as we have been describing, the simple force of water acting in mass on the surface of gently inclined and regular strata of chalk and oolite is sufficient for the effects produced; but in other cases, more especially in mountain districts, (where the greatest disturbances appear generally to have taken place,) the original form in which the strata were deposited, the subsequent convulsions to which they have been exposed, and the fractures, elevations, and subsidences which have affected them, have contributed to produce valleys of various kinds on the surface of the earth, before it was submitted to that last catastrophe of an universal deluge which has finally modified them all.

Fig.1.   **Pl.2.**

Fig.2.

4 feet

Section of the Cave

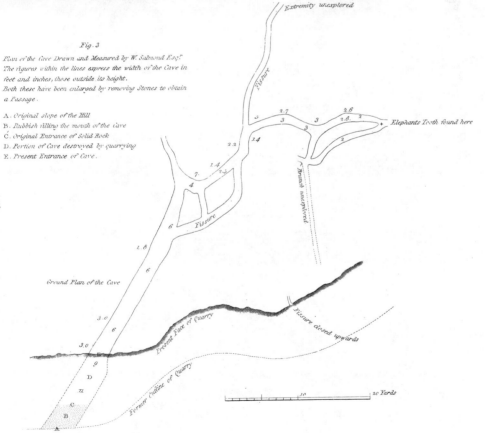

Extremity unexplored

Fig.3

Plan of the Cave Drawn and Measured by W. Salmond Esq.r
The figures within the lines express the width of the Cave in
feet and inches, those outside its height.
Both these have been enlarged by removing Stones to obtain
a Passage.

A . Original slope of the Hill
B . Rubbish filling the mouth of the Cave
C . Original Entrance of Solid Rock
D . Portion of Cave destroyed by quarrying
E . Present Entrance of Cave.

Fissure

Elephants Tooth found here

Branch unexplored

Fissure

Ground Plan of the Cave

Present Face of Quarry

Fissure closed upwards

Former Outline of Quarry

10   20 Yards

W. Buckland \ del.
T. Webster /

J. Basire sc.

Pl. 1.

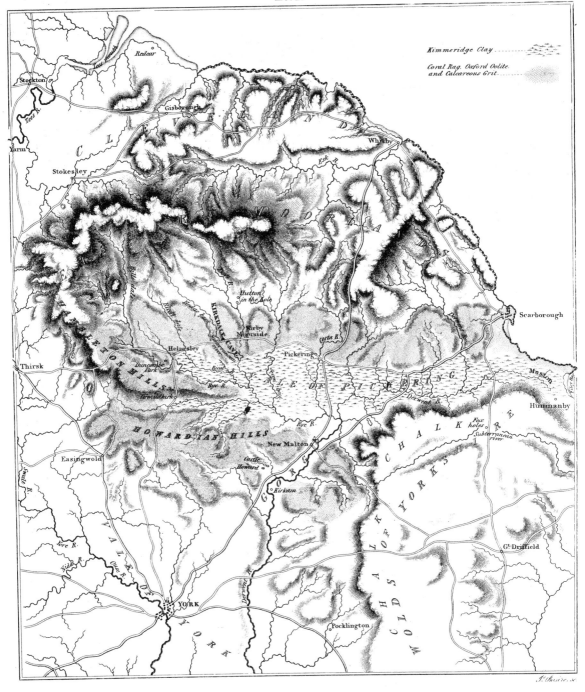

J. Basire sc.

EXPLANATION OF THE PLATES.

All the Drawings are of the natural size, except where it is expressed to the contrary.

PLATE I.

Map of the country adjacent to the cave at Kirkdale, showing the entire drainage of the vale of Pickering to be effected through the gorge at Malton, the stoppage of which would at once convert it into an inland lake.

PLATE II.

Fig. 1. View of the mouth of the cave at Kirkdale, in the face of a quarry, near the brow of a low hill.

Fig. 2. Section of the cave before the mud had been disturbed.

A. Stratum of mud covering the floor of the cave to the depth of one foot, and concealing the bones.

B. Stalagmite incrusting some of the bones, and formed before the mud was introduced.

C. C. Stalagmite formed since the introduction of the mud, and spreading horizontally over its surface.

D. Insulated stalagmite on the surface of the mud.

E. E. Stalactites hanging from the roof above the stalagmites.

Fig. 3. Ground plan of the cave, by W. Salmond, Esq. showing its extent, ramifications, and the fissures by which it is intersected.

Plate III.

1. Portion of the left upper jaw of the modern hyæna from the Cape.

2. Inside view of No. 1.

3. Analogous portion of the left upper jaw of the fossil hyæna from Kirkdale.

4. Inside view of No. 3, with the tooth of a water-rat adhering by stalagmite to a broken portion of the palate.

5. Fragment from Kirkdale, showing five incisor teeth of the upper jaw, much worn down, and the inside of the palate.

Plate IV.

1. Outside view of the right lower jaw of the modern Cape hyæna.

2. Analogous portion of lower jaw of the Kirkdale hyæna, being nearly one-third larger.

3. Inside view of No. 2.

Plate V.

1. Fragment of the right lower jaw of an hyæna, showing the convex surface of the jaw and its teeth, that lay uppermost in the den, to be deeply worn by friction, and to have received a polish. The enamel and one-third of the substance of the teeth and bone on this side have been worn away.

2. Concave surface of No. 1, having no marks of friction, polish, or decay: the enamel on this side of the teeth is perfect and unchanged.

3. Fragment of the right lower jaw of a young hyæna, having the convex surface only polished as in No. 1; and showing the cavities in which the second set of teeth were rising to succeed the first set; one of which, the posterior molar tooth, still remains in its place,

Pl. 3.

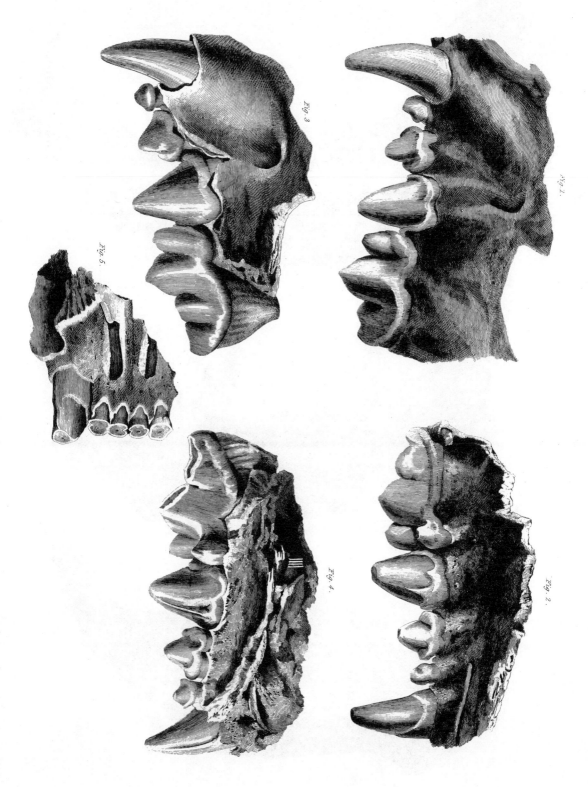

Fig. 3.

Fig. 1.

Fig. 5.

Fig. 4.

Fig. 2.

T. Webster del.

J.ᵉ Basire sc.

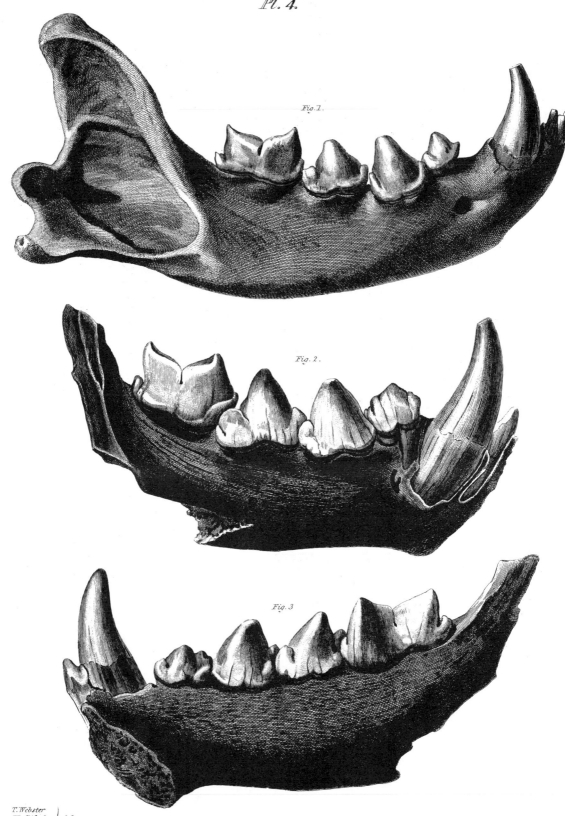

Pl. 4.

Fig.1.

Fig. 2.

Fig. 3.

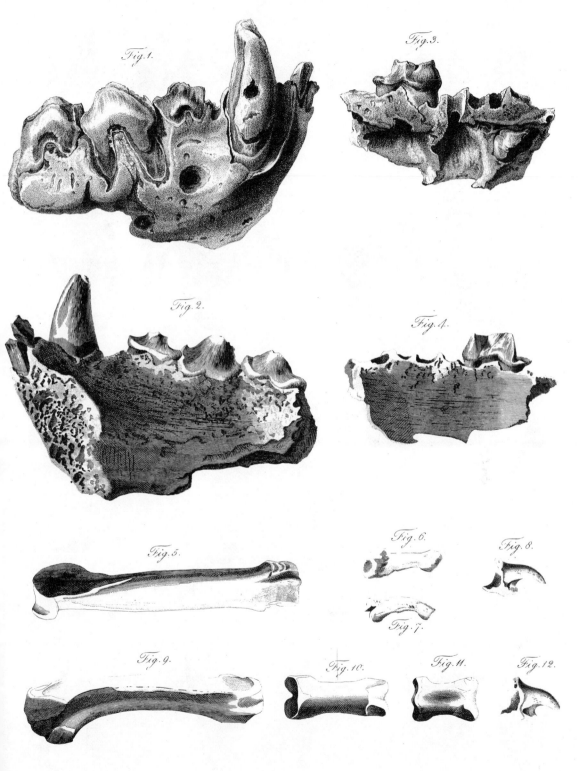

Pl. 5.

Fig. 1.

Fig. 3.

Fig. 2.

Fig. 4.

Fig. 5.

Fig. 6.

Fig. 8.

Fig. 7.

Fig. 9.

Fig. 10.

Fig. 11.

Fig. 12.

F. Duncombe del.
T. Webster

J. Basire sc.

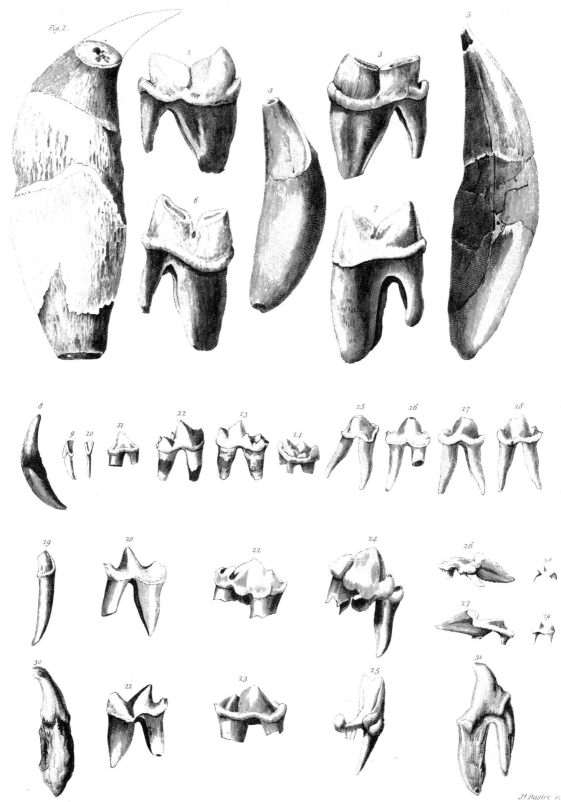

having its enamel on one side only worn away, as in the teeth of No. 1. See also Plate XIII. fig. 3, 4. and Plate XXIII. fig. 9, 10.

4. Inside or concave surface of No. 3; this side has suffered no friction or polish, and the enamel of the tooth is perfect and fresh as in No. 2.

5. Metatarsal bone of hyæna.

6, 7. Phalanges of the toe of an animal not ascertained.

8. Claw bone of the toe of an hyæna.

9, 10, 11. Metacarpal bone and two phalanges of the toe of hyæna.

12. Claw bone of the toe of an hyæna.

PLATE VI.

1. Canine tooth or tusk of a bear, (apparently Ursus spelæus).

2. Inside view of posterior molar tooth of the lower jaw on the left side of hyæna.

3. Outside view of No. 2.

4. Largest canine tooth or tusk of hyæna found at Kirkdale.

5. Tusk of an animal of the tiger kind; four teeth only of this sort have been found at Kirkdale.

6. Outside view of right posterior molar tooth of the lower jaw of a tiger; four of these teeth also are all that have yet been noticed at Kirkdale.

7. Inside view of No. 6. On comparing 6 and 7 with 2 and 3 it will be observed that in 6 and 7, the angle near the middle part of the crown is less obtuse than in 2 and 3, and that the two lobes which project at the base of the crown of 2 and 3 are wanting in 6 and 7.

8. Tusk of fox.

9. Incisor tooth of fox.

10. Inside view of No. 9.

11. Small molar tooth of fox.

12. Great molar tooth of the right lower jaw of fox; outside view.

13. Inside view of No. 12.

14. Penultima of upper jaw, right side, of fox.

15. ⎫
16. ⎭ Molar teeth of the first set of a young hyæna.

17. Inside view.

18. Outside view of No. 17.

19. Canine tooth of a young hyæna.

20, 21. Posterior molar tooth of the lower jaw of a young hyæna, much worn.

22, 23, 24, 25. Outside and inside views of two molar teeth of the upper jaw of a young hyæna: they are all extremely thin, and have deep furrows worn on them. They may be seen in the jaw at Plate XIII. No. 3, 4.

26, 27. Posterior molar tooth of the upper jaw of a young hyæna.

28, 29. Posterior tooth and penultima of a weasel, left upper jaw (twice the natural size).

30, 31. Side and front views of the same tooth, probably a diseased molar tooth of an hyæna.

PLATE VII.

1. Small molar tooth of a very young elephant, being the average size of those found in the den.

2. Fragment of a still younger elephant's tooth.

3. Molar tooth of upper jaw of rhinoceros.

4. Inside view of molar tooth of lower jaw of rhinoceros.

5. Crown of No. 4, as seen from above.

6 Outside view of No. 4.

7. Molar tooth of the upper jaw of a horse.

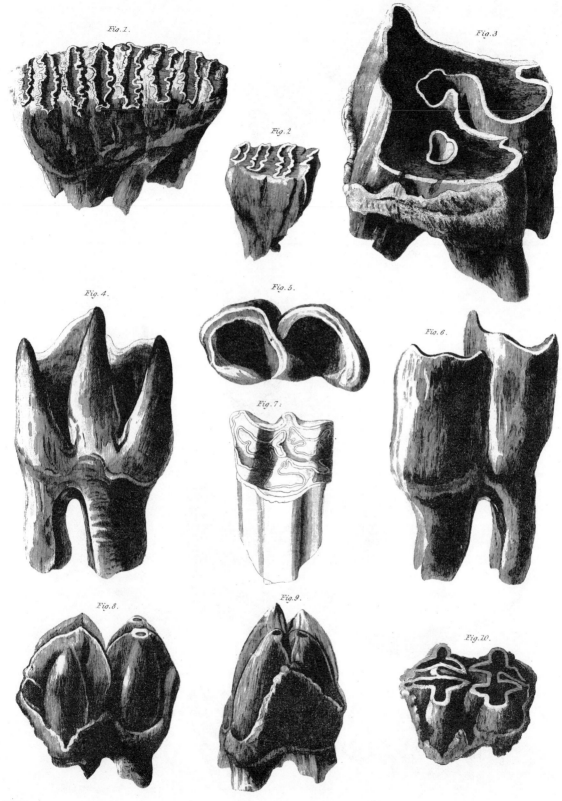

Fig. 1.

Fig. 2.

Fig. 3

Fig. 4.

Fig. 5.

Fig. 6.

Fig. 7.

Fig. 8.

Fig. 9.

Fig. 10.

M. Morland
W. Clift &   del.
T. Webster

J. Basire sc.

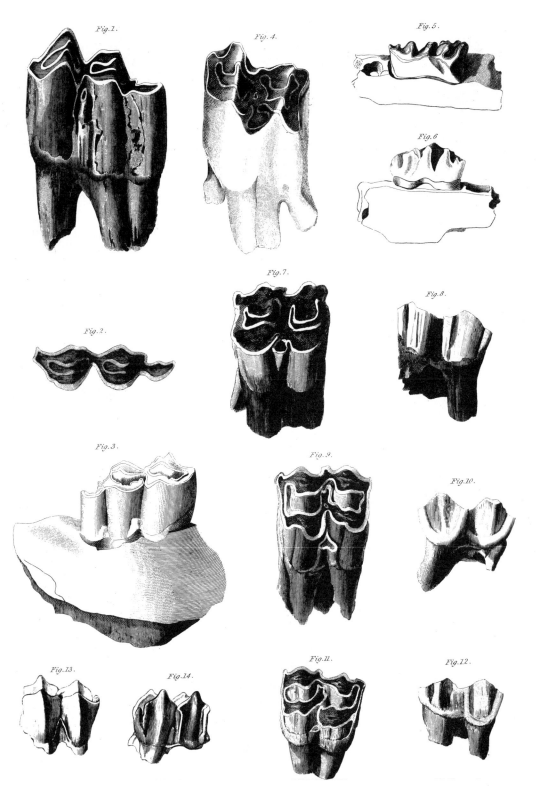

Fig.1.

Fig.4.

Fig.5.

Fig.6.

Fig.7.

Fig.8.

Fig.2.

Fig.3.

Fig.9.

Fig.10.

Fig.13.

Fig.14.

Fig.11.

Fig.12.

M. Morland.
E. Duncombe. del.
T. Webster.

J. Basire sc.

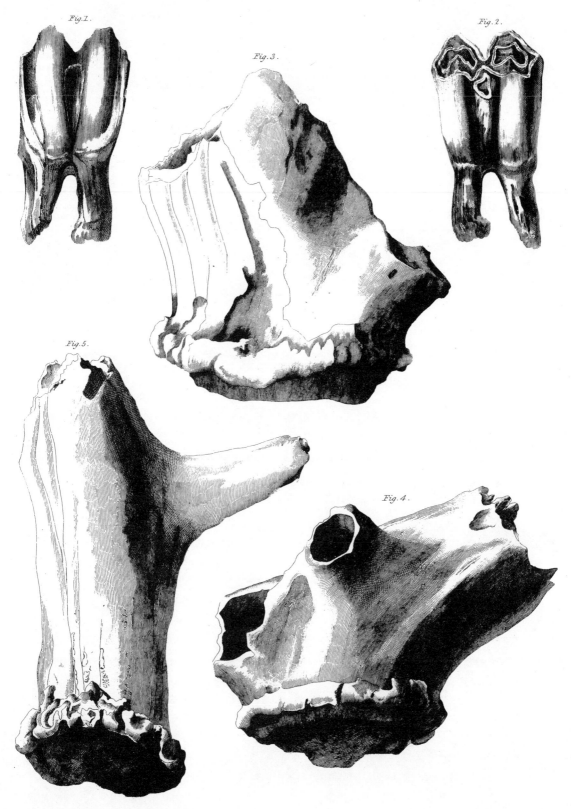

Fig.1.

Fig.2.

Fig.3.

Fig.5.

Fig.4.

E. Duncombe del.

J.^s Basire sc.

8.⎫ Two views of a molar tooth of hippopotamus not yet worn
9.⎭   down.

10. Molar tooth of hippopotamus, having the summits of the crown worn down.

Plate VIII.

1. Posterior molar tooth of the lower jaw of an ox.

2. Crown of No. 1.

3. Posterior molar tooth of the right lower jaw of a species of deer.

4. Molar tooth of the upper jaw of an ox.

5. Molar tooth of the lower jaw of a calf.

6. Side view of No. 5.

7. Molar tooth of the upper jaw of an ox.

8. Outside view of No. 7.

9. Molar tooth of the upper jaw of a very large species of deer, equalling in size the largest elk, but differing in form.

10. Outside view of No. 9.

11. Molar tooth of the upper jaw of a second species of deer, equalling in size the largest red deer.

12. Outside view of No. 11.

13.⎫ Inside and outside views of a rising molar tooth of a third
14.⎭   species of deer, of the size of a large fallow deer.

Plate IX.

1. Outside view of a molar tooth of the lower jaw of a large species of deer.

2. Inside view of No. 1.

3. Base of the horn of a large deer, measuring nine inches and three-quarters in circumference, which corresponds exactly in

size with that of a very large English red deer in the Anatomy School at Oxford.

4. Base of a horn similar to No. 1, having two antlers near its lower extremity, and measuring seven inches and three-quarters in circumference.

5. Base of a deer's horn, having the lowest antler at the distance of three inches and a half from the lower extremity.

PLATE X.

1. Coronary bone of a horse.

2. First phalangal bone of a very large ox; seen laterally.

3. Under side of No. 2.

4, 5. Astragalus of a large ox: two different sides of the same bone.

6. Album græcum, showing a small sphere adhering to the larger one, and an indentation of the sides of both by pressure from a third sphere.

7. Astragalus of hyæna.

8. Side view of No 7.

9. Astragalus of fox.

10. Side view of No. 9.

11. Astragalus of water-rat.

12. Os calcis of water-rat.

13. Os calcis of fox.

14. Os calcis of rabbit.

15, 16. Internal metatarsal bone of rabbit, having lost the epiphysis.

17, 18. Metatarsal bone of rabbit, retaining the epiphysis at the lower extremity.

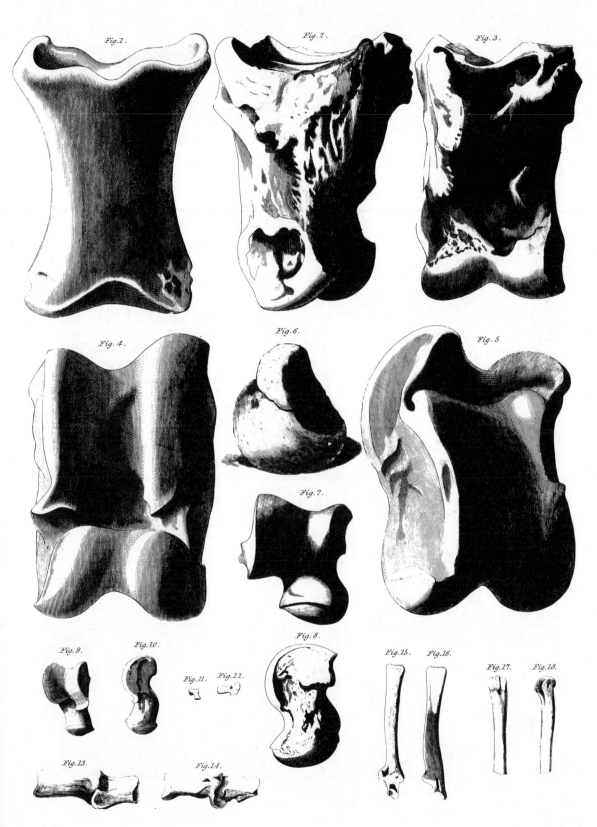

Fig. 1. Fig. 2. Fig. 3.

Fig. 4. Fig. 6. Fig. 5.

Fig. 7.

Fig. 9. Fig. 10. Fig. 11. Fig. 12. Fig. 15. Fig. 16. Fig. 17. Fig. 18.

Fig. 8.

Fig. 13. Fig. 14.

T. Webster del.

J.ᵗ Basire sc.

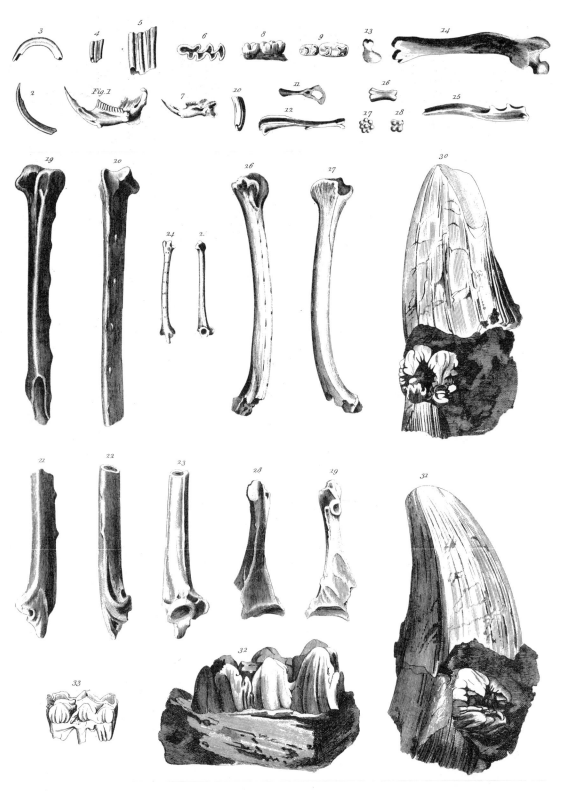

Pl. II.

M. Morland.
T. Webster,   del.t
H. O'Neil.

Jt Basire sc.

PLATE XI.

The specimens from 1 to 29, inclusive, are all from Kirkdale.

1. Lower jaw of a water-rat.
2. Lower incisor tooth of No. 1.
3. Upper incisor of water-rat.
4. Anterior molar tooth of lower jaw of water-rat.
5. No. 4, magnified.
6. Crown of No. 4, magnified.
7. Jaw of a mouse.

8 and 9. Teeth of No. 7, magnified four times.

10. Anterior molar tooth of the upper jaw of a rabbit.
11. Os innominatum of a young water-rat.
12. Tibia of a water-rat.
13. Lower epiphysis of femur of water-rat, twice magnified.
14. Femur of water-rat, twice magnified.
15. Ulnar of water-rat.
16. Tail vertebra of water-rat.
17. Anterior extremity of No. 16.
18. Posterior extremity of No. 16.
19. Right ulna of a raven; anterior extremity.
20. Outside view of No. 19, showing the points of attachment of the quill feathers.
21. Right ulna of a raven, showing the other extremity of No. 19.
22, 23. Other views of No. 20, showing the cavity to be nearly filled with stalagmite.
24. Right ulna of a lark, showing the attachments of the quill feathers.
25. Inside view of No. 24.
26. Left ulna of a very large species of pigeon.

27. Inside view of No. 26.

28. Right coracoid process of the scapula of a small species of duck or widgeon.

29. Inside view of No. 28.

30. Tusk of the upper jaw of a large hog, polished obliquely near its apex, and having a molar tooth of hog adhering to it, near its base, by an ochreous crust, from Hutton, in Mendip.

31. View of the opposite side of No. 30.

32. Large molar tooth of hog in a fragment of the lower jaw, slightly incrusted with ochre, from Hutton.

33. Small molar tooth of hog, from Hutton.

PLATE XII.

Lower jaw nearly entire of a very old hyæna, found with the bones of elephant, rhinoceros, horse, ox, &c. in diluvium, at Lawford, near Rugby, in Warwickshire. The coronary part of all the teeth is nearly worn off, and on the worn surface of the two hindmost there are deep furrows; all the surfaces are highly polished, and even have a brilliant lustre; seven teeth only remain, the animal having worn out nine from its lower jaw alone, viz. six incisors, the left canine or tusk, and two anterior molars. Traces of the root of the right anterior molar are still visible in their proper place: the sockets of all the other lost teeth have been either removed by absorption, or filled up with bone.

It should be observed, that this specimen, and the humerus and ulna (Plate XIII. 1, 2.), are not in the least degree mangled or broken like those from the den at Kirkdale, being derived probably from one of the last hyænas that were drowned by the diluvian waters, together with the other animals whose bones are found with them equally perfect, and free from such marks of violence as occur on all the bones of whatever kind discovered at Kirkdale.

Pl. 12.

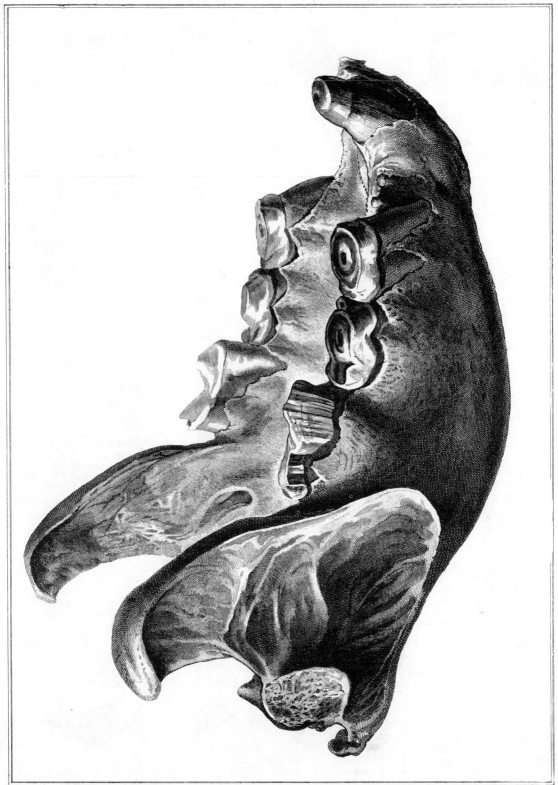

T. Webster, delt.

Jas Basire, sculpt.

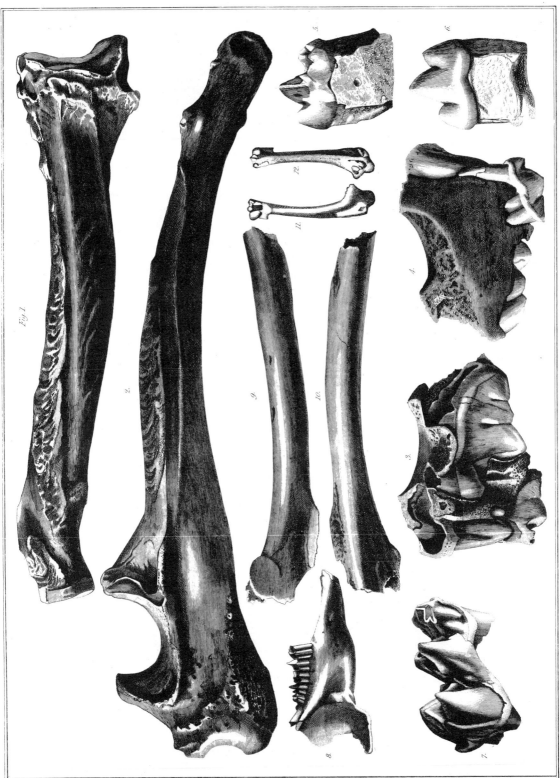

PLATE XIII.

1. Radius of a very old hyæna, found at Lawford, in the diluvium, with the jaw, in Plate XII, and probably from the same individual.

2. Ulna fitting No. 1, and found together with it. Both these bones, like the jaw, have no marks of gnawing or any violence in them, and appear to be from the same individual. Nos. 1 and 2 are reduced one-tenth.

3. Fragment of the right upper jaw of a young hyæna, showing the second set of teeth advancing beneath the first. This interesting specimen belongs to Archdeacon Wrangham, on whose property at Kirkdale the cavern stands.

4. Outside view of No. 3; compare with this, the teeth 22, 23, 24, 25, at Plate VI.

5. Inside view of the great molar tooth of the right lower jaw of a wolf found at Kirkdale, by Mr. Salmond.

6. Outside view of No. 5.

7. Small molar tooth of an hippopotamus from Kirkdale.

8. Part of the lower jaw of a hare or very large rabbit, from Kirkdale. Nos. 7 and 8 are from drawings, by Miss Duncombe.

9. Humerus of a bird, apparently a goose, found at Lawford; with Nos. 1 and 2, and with No. 1, Plate XII.

10. Outside view of No. 9. This bone is the only example I know of the remains of birds being noticed in the diluvium of England, excepting those at Kirkdale.

11. Humerus of a bird, apparently a snipe, from Kirkdale. I had not compared this bone with any recent skeletons, at the time when page 15 was printed.

12. Inside view of No. 11.

Plate XIV.

Vertical section of the great cave of Scharzfeld, on the west border of the Hartz; drawn from a sketch made on the spot by Professor Buckland, A. D. 1822.

A. Fissure in the surface of the land by which we descend into the great chamber B.

B. Portion of the main chamber, which extends into the hill to at least three times the length here represented; its roof is hung with clusters of stalactite.

C. Crust of stalagmite, restored to perhaps a greater degree than that in which it probably existed on the floor before it had been dis_turbed in search of bones.

D. Bed of brown earth or diluvial loam, spread over the actual floor of the cave, and interspersed with angular fragments and rounded pebbles of limestone, and a few teeth and bones.

E. Artificial excavation in this brown earth, down to the limestone of the actual floor.

F. Artificial excavation through brown earth into the under-vault G.

G. Under-vault filled completely with diluvium similar to D, but much more abundantly loaded with bones.

H. Artificial vault excavated in G, in search of bones, which are seen forming part of its roof and sides, as well as of its floor.

I. Under-vault, filled with the same diluvium and bones as G, and not yet disturbed.

K. Passage, communicating from G to I, and also filled in the same manner.

Pl. 14.

Drawn by T.Webster from a Sketch by W.Buckland 1822.

Printed by c.Hullmandel.   G.Scharf.Lithog.

VERTICAL SECTION OF THE CAVE OF SCHARZFELD.

Pl. 15.

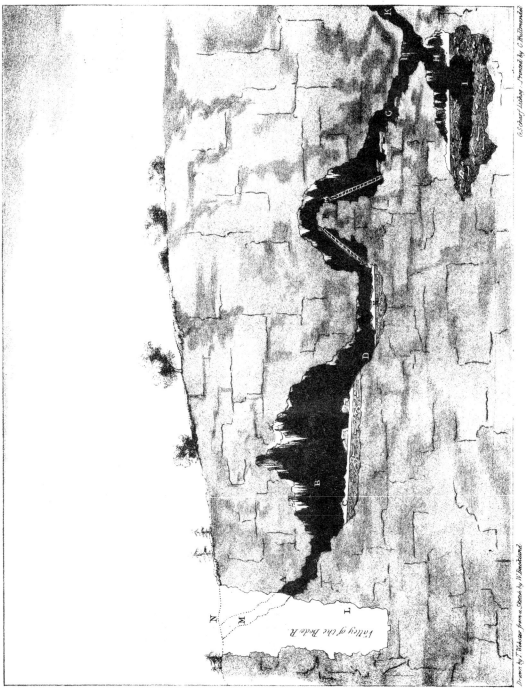

VERTICAL SECTION OF THE CAVE OF BAUMANS HÖHLE IN THE HARTZ.

PLATE XV.

Vertical section of the cave of Bauman's Höhle, in the Hartz, in the cliff on the left side of the gorge of the river Bode.

A. Present entrance, descending from the perpendicular cliff to the large chamber B.

B. Grand chamber, from 40 to 50 feet in diameter, and having stalactite on various parts of its roof.

C. Floor of the great chamber, covered with a bed of diluvial loam, containing a few bones and pebbles, and having on its surface some residuary fragments of a crust of stalagmite and large blocks of limestone, marked o. o.

D. Inclined passage communicating from B. to the lower vault, E.

E. Lower vault, containing beneath a thick crust of stalagmite a considerable accumulation of loam and large pebbles, through both of which are dispersed great quantities of bones of bears.

F. Rugged mass of rock, ascending about 20 feet, and dividing the vault E. from the inner chambers G. H.

G. Long irregular chamber conducting from F. to the lower cavern H.

H. Lower cavern, having its roof much hung with stalactite, and on its floor a considerable crust of stalagmite.

I. Artificial vault dug through the crust of stalagmite into a subjacent mass of loam, pebbles, and bones, accumulated to the thickness of several feet.

K. An unexplored passage, ascending from the roof of the cave H.

L. Deep gorge, being a valley of denudation, flanked on each side by steep cliffs of limestone.

M. Supposed continuation of the cave A. to the antediluvian surface N.

N. Supposed surface of limestone as it existed before the excavation of the valley L.

o. o. Large block of limestone laid irregularly in the chamber B., and apparently fallen from the roof.

Plate XVI.

Section of the cave of Biel's Höhle, nearly opposite that of Bauman's
Höhle, and on the right side of the gorge of the Bode.

A. Small hole of entrance in the side of the cliff, by which we descend into the suite of irregular chambers that compose this cave.

B. Bottom of numerous hollows or basins that occur along the course of the cave, and are uniformly covered with a deep bed of mud and sand, over which is spread a crust of stalagmite.

C. Tubular cavities and fissures that ascend from various parts of the cave towards the surface, and by which the mud and sand were probably drifted in.

D. Irregular rocky masses that form large pinnacles between the basins B., and have cavities on their summits filled also with a deep sediment of mud and sand, the surface of which is sealed over with a crust of stalagmite, H.

E. Gorge or narrow valley of the Bode river, flanked on both sides by precipitous crags of transition limestone.

F. Supposed continuation of the mouth A. to the surface, as it probably existed before the excavation of the valley.

G. Supposed surface of the limestone before the formation of the subjacent valley.

H. Crust of stalagmite covering the diluvial mud and sand both in the hollows B., and on the pinnacles D.

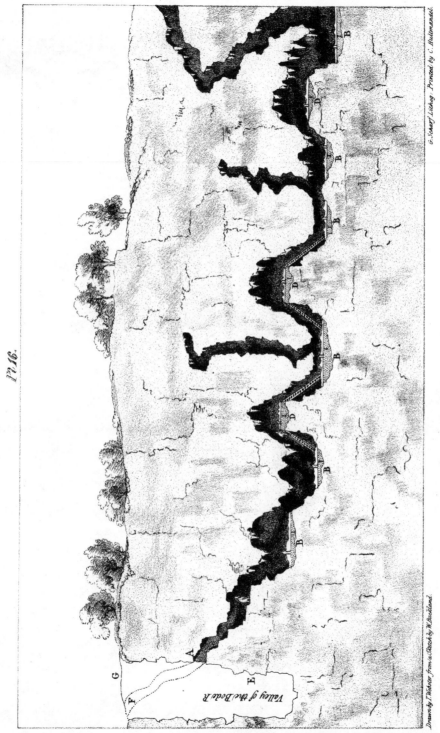

Pl. 16.

Drawn by J.Webster from a Sketch by W.Buckland.

G.Scharf Lithog. Printed by C.Hullmandel.

SECTION OF THE CAVE OF BIELS HÖHLE IN THE HARTZ.

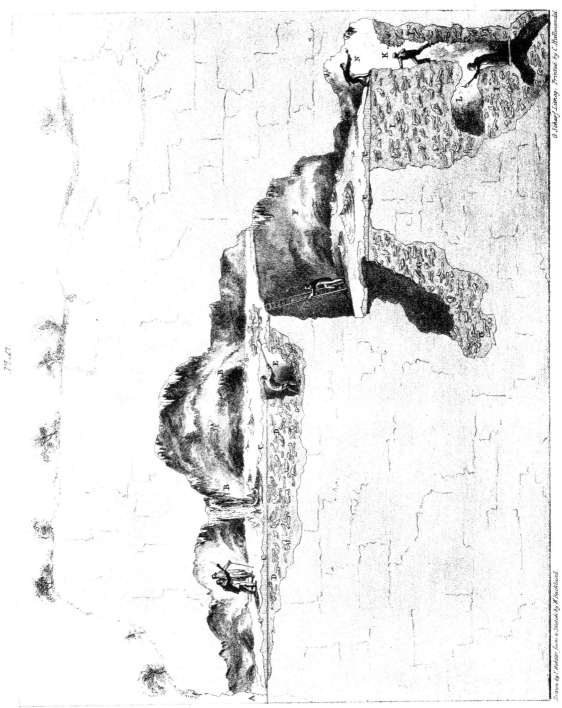

Fig. 47.

Drawn by T. Webster from a Sketch by W. Buckland.

G. Scharf Lithog. Printed by C. Hullmandel.

VERTICAL SECTION OF THE CAVERN AT GAILENREUTH IN FRANCONIA.

Plate XVII.

Vertical section of the cave of Gailenreuth, in Franconia, from sketches made by Professor Buckland in 1816 and 1822.

A. Entrance passage, varying from six to ten feet in height, terminating externally in a steep cliff, and internally expanding itself into the large chamber B.

B. Large chamber, having much stalactite on its roof, and still more stalagmite on its floor. In the centre is a pillar of these substances, uniting the roof and floor.

C. Crust of stalagmite, still perfect over great part of the floor of B., but much destroyed on that of the lower chamber F.

D. Bed of diluvial loam, mixed with pebbles, angular fragments of limestone, bones and teeth: the bones are not so abundant as in the lower mass G. I.

E. Hole excavated in the mass D. for the purpose of extracting bones: fragments of these bones lie loosely scattered on the surface of the crust C., and mixed with bones of modern animals, with ashes and charcoal from fires made to illuminate the cave, and with common soil brought in from the external surface.

F. Second chamber, separated from B. by a perpendicular precipice, and having probably other less steep communications with it. The stalagmite of the floor C. is represented as restored to the state in which it probably existed before it had been disturbed by digging.

G. Enormous mass of bones lying in loose earth in a deep natural cavern, which descends laterally from the chamber F.

H. Upper part of the cavern which contains the bones G.

I. Mass of bones, 25 feet deep, mixed with pebbles and loam, and cemented by stalagmite into a strong osseous breccia.

K. Well sunk 25 feet deep in I., for the purpose of extracting bones.

K. K. Cavities excavated at the bottom of K., but not reaching through the breccia to the natural limestone rock.

L. Oven-shaped cavity dug in the side of I. in search of bones and skulls.

M. Low passage connecting the chamber F. with the smaller chamber N.

N. Small innermost chamber, in the floor of which is sunk the well K.   This must originally have been the roof of a deep cave, which has been filled up by the mass I. I.

PLATE XVIII.

View of the narrow valley or gorge of the Esbach river, which falls into the Weissent a little above Muggendorf.

A. Ruins of the castle of Rabenstein, on the edge of a cliff about 100 feet high, on the right bank of the Esbach.

B. Chapel of Klaustein, standing immediately over the cave C., which I have called the cave of Rabenstein: it also bears the name of Klaustein.

C. Mouth of the cavern, leading to a large chamber, which has many side vaults and lateral communications, some of which probably pass upwards to the surface.   This cave contains few bones, but much mud and stalagmite.

D. Channel of the Esbach, a very small river which descends by this gorge to join the Weissent.   The gorge in its narrowest part is not 50 yards broad.

E. Mouth of the cave of Kühloch, in the lowest part of the cliff, on the left flank of the gorge opposite the castle.   This mouth must

Pl. 18.

VIEW OF THE MOUTHS OF THREE CAVES in the Cliffs that flank THE GORGE OF THE ESBACH RIVER
BY THE CASTLE OF RABENSTEIN IN FRANCONIA

Pl.19.

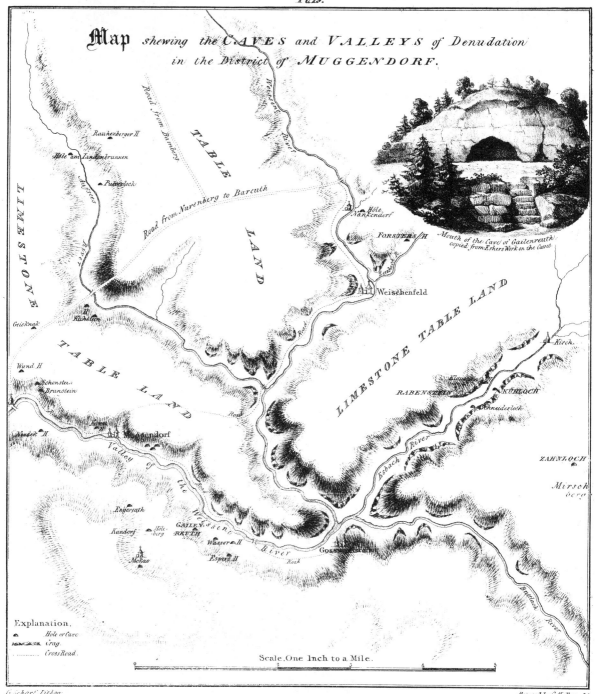

Map shewing the CAVES and VALLEYS of Denudation in the District of MUGGENDORF.

Mouth of the Cave of Gailenreuth copied from Eshers Work on the Caves.

TABLE LAND

LIMESTONE TABLE LAND

Road from Bamberg

Road from Nurenberg to Bareuth

Rauhenberger H

Hole am Lindenbrunnen

Putzerloch

Aufsees River

Hole Nankendorf

FORSTERS H

Weischenfeld

Geisknäk

Kühstein

TABLE LAND

Wund H

Schonstein

Branstein

Mudek H

Wiesent River

Esbach River

RABENSTEIN

KÜHLOCH

Schneiderloch

Kirch

ZAHNLOCH

Mirsch berg

Valley of the Weissene River

Engereuth

Kandorf

Hole berg

GAILENRKUTH

Wasser H

McKas

Espera H

Goss

Road

Büttloch River

LIMESTONE

TABLE LAND

Explanation.
Hole or Cave.
Crag.
Cross Road.

Scale, One Inch to a Mile.

G. Scharf Lithog.

Printed by C. Hullmandel.

Pl. 20.

Drawn by T. Webster from a Sketch by Prof. Buckland.

G. Scharf Lithog.

SECTION OF THE CAVE IN THE DREAM LEAD MINE NEAR WIRKSWORTH, DERBYSHIRE. 1822.

have been included in solid rock till the valley had been cut down nearly to its present depth.

F. Entrance to the cavern by a lofty vault leading to the cave C. The roof and floor of this vault are inclined downwards at a considerable angle.

G. Section of the interior of the great cave, closed on every side with solid rock except at F., and having its floor H. buried under a deep bed of black animal earth and bones. The section is represented by dotted lines marked on the surface of the cliff.

H. Actual floor of the cave, beneath the black earth and bones.

I. Mouth of the cave of Schneiderloch, in the cliffs a little below Kühloch, and also containing bones.

Plate XIX.

Map of the district round Muggendorf, in Franconia, showing the manner in which the country is intersected by deep valleys of denudation, and the present mouths of the caves exposed in the cliffs that flank these valleys, though not exclusively confined to them. This map is copied from that in Goldfuss's Pocket-book on the Environs of Muggendorf.

The Vignette, giving a view of the mouth of the cave of Gailenreuth, with a fissure close adjacent to it, is copied from an engraving in Esper's account of the caves of this district.

Plate XX.

Vertical section of the cave discovered in the Dream lead mine at Callow, near Wirksworth, Derbyshire, in December, 1822.

A. Shaft sunk perpendicularly downwards 60 feet, through a solid vein containing lead.

B. Supposed continuation of the lead vein below the floor of the cave.

N N

C. Cave in the state it was when visited by the author in January, 1823, excepting that a large number of the rhinoceros bones had been extracted.

D. Fissure laid open by the subsiding of the materials that had filled it, into the cave C.; the face of this fissure, under the plumb-line, is rubbed and scratched as if by descending masses of stone.

E. Subsided mass of loose stones and argillaceous loam that had filled the cave to its roof, and the fissure to its surface, before the cave was penetrated, and its contents in part extracted by the shaft A.

F. Bones of ox and deer, and horns of deer found near those of the rhinoceros.

G. Skeleton of rhinoceros restored to the state in which it probably lay before its matrix had been disturbed by subsiding towards the shaft.

H. Solid limestone of Derbyshire, containing the now open fissure D. and the shaft A., and intersected by numerous lead veins.

I. Surface of the fissure, which was entirely level, and overgrown with grass, till its contents began to subside into the cave C.

Plate XXI. Fig. 1.

Vertical section of the cave of Goat Hole at Paviland, in the sea cliff, 15 miles west of Swansea, in Glamorganshire.

A. Mouth of the cave, at the base of a nearly vertical cliff, facing the sea, and accessible only at low water, except by dangerous climbing.

B. Inner extremity of the cave, where it becomes so small, that a dog only can go further, and apparently ending at a short distance within B.

C. Body of the cave. Its length from A. to B. is about 60 feet, the

Pl. 21.

Ground Plan of the Cave.

SECTION OF THE CAVE CALLED GOAT HOLE.
In the Sea Cliffs 15 Miles West of Swansea.

Drawn by T. Webster from a Sketch by Prof. Buckland.

C. Scharf Lith. Printed by C. Hullmandel.

breadth from C. to D. (in the plan, fig. 2,) is about 20, the height of the cave from 25 to 30 feet.

D. (In the section) irregular chimney-like aperture, ascending from the roof of the cave, and terminating in the nearly perpendicular cliff at K.; it is too small for the entire carcase of an elephant to have passed down through it.

E. Bottom of the cave, to which the sea water never reaches; this part is covered over with a loose mass of argillaceous loam and fragments of limestone, of diluvial origin, about six feet deep, which has been much disturbed by ancient diggings, and through which are dispersed the bones and teeth. The elephant's head, and human skeleton, are marked in the spot in which they were actually found.

F. Mass of the same materials as E., but less disturbed, and overhanging E. with a small cliff, five feet high, in which were found two elephant's teeth. This mass, though less disturbed than E., has been dug over before, and extends into the small hole within B.; it contains dispersed through it, particularly at B., recent sea shells and pebbles: at this place also it is firmly united by stalagmite, which rarely occurs in any other part of the cave.

G. Loose sea pebbles, strewed in small quantity over the floor of the cave near its mouth, and washed up only by the waves of the highest storms.

H. Rock basins, three feet deep, produced by friction of the large pebbles, which still lie in them.

I. Naked limestone of the floor of the cave, forming the line within which the waves appear never to enter, and separating the sea pebbles without, from the diluvial loam and angular fragments that form the loose breccia within it.

K. Upper termination of the chimney-shaped aperture in the face of the naked cliff.

L. Ledges and hollows in the aperture K, on each of which is lodged about a foot of loose fine earth, that seems to be accumulated from dust drifted by the wind, and is full of minute land shells, and the bones of small animals, apparently brought hither by hawks and sea gulls, e. g. moles, water-rats, field-mice, small birds, and fish.

PLATE XXII.

1. Lower portion of the horn of a small deer, apparently a cast horn separated by necrosis; found in the Goat's Hole Cave at Paviland.

2. Upper extremity of another horn found with No. 1, apparently of the same species; it is very flat and thin.

3. Lower extremity of a horn, still adhering to the skull, found with the rhinoceros in the cave near Wirksworth, 1822. Near it were several cylindrical portions of the shaft of similar horns, nearly of the same diameter as this, having their surface very smooth.

4. Portion of a flat and palmated horn found with No. 3. The scale of 1, 2, 3, and 4, is half the natural size.

5. Head of an hippopotamus, copied from p. 185 of Lee's History of Lancashire (fol. Oxon, 1700). The only account given of it is that it was dug up under a moss in Lancashire.

6. Outside view of the right tusk of the upper jaw of an animal of the tiger kind, found by Mr. Cottle, of Bristol, A. D. 1822, with the bones already described, in the cave at Oreston, near Plymouth.

7. Inside view of No. 6.

PLATE XXIII.

1. Residuary part of the lower extremity of the tibia of an ox, which I saw given entire to a Cape hyæna in Mr. Wombwell's travelling collection at Oxford, in December, 1822: marks of the teeth are

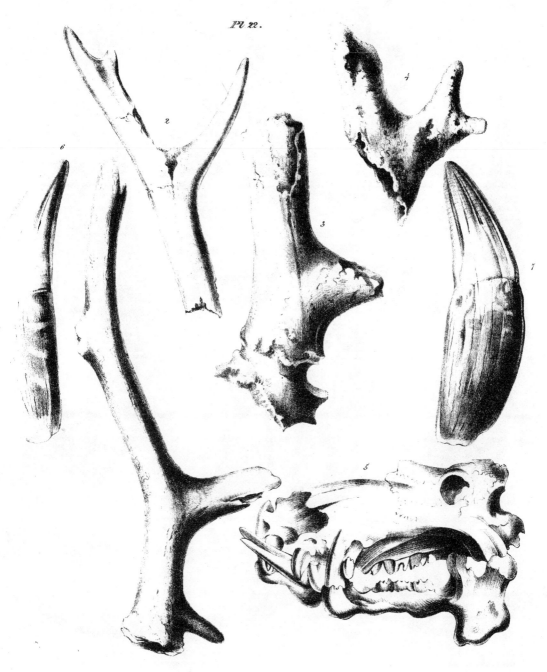

Pl 22.

G Scharf lithog.

Printed by C Hullmandel

Pl. 23.

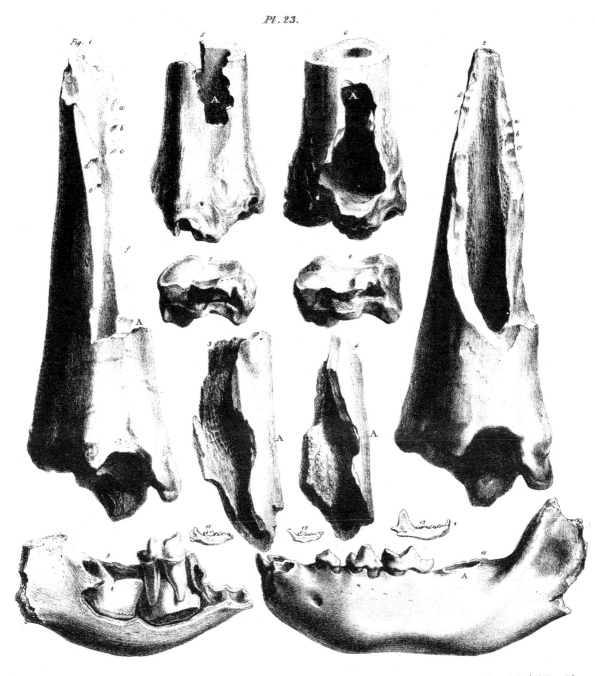

Drawn from Nature & on Stone by G. Scharf

Printed by C. Hullmandel.

RECENT AND ANCIENT MARKS OF THE TEETH OF HYÆNAS,
JAWS OF WEASEL AND OF YOUNG HYÆNAS.

distinctly visible at a, b, c, d, e, f. See note at p. 37, describing the hyæna's manner of breaking and partly devouring this bone. Nos. 1, 2, 3, and 4, are reduced one-third.

2. Fragment of a similar tibia from Kirkdale Cave, broken nearly in the same manner as No. 1, and bearing similar marks of teeth at a, b, c, d, e: in the recent, as in the antediluvian specimen, the lower condyle has, from its hardness, been left unbroken.

3. Splinter from another bone broken by the hyæna at Oxford: the cavity at A. was produced by the hyæna's tooth.

4. Similar splinter, bearing a similar cavity A., from the cave at Kirkdale, and partially incrusted with stalagmite.

5. Inside view of the lower extremity of the recent specimen, No. 1, in which the hole A. was produced by the hyæna's bite. Nos. 5 and 6 are reduced nearly one-half.

6. Lower extremity of another tibia from Kirkdale, in which the form of the cavity A. resembles that in No. 5.

7. Scaphoid bone of the left carpus of an ox, which, with the other component bones of the carpus, lay all night untouched in the hyæna's cage at Oxford.

8. Similar bone from Kirkdale, equally untouched.

The above specimens go far to explain the fact of the abundance (in excess) at Kirkdale of such solid and marrowless bones, and fragments of bone; and of the absence of the softer portions analogous to those which were devoured by the hyæna in Oxford.

9. Fragment of the lower jaw of a young hyæna from Plymouth, exhibiting the posterior molar tooth of the first set about to be shed, and two of the permanent teeth rising in the jaw beneath; from a drawing by Mr. Clift.

10. Jaw of a young hyæna from Kirkdale, belonging to Mr. Salmond, showing it to have but three deciduous molar teeth in the

lower jaw, whilst the number of permanent molar teeth in this same jaw is four: at A. the posterior tooth of the second set is in the act of rising through the bone, but not yet protruded.

11, 12, 13. Jaws of weasels from Kirkdale, belonging to Mr. Salmond.

PLATE XXIV.

Copy of a drawing by Schröder, published by Mr. Bieling, of a remarkable mass of remains discovered in a bed of diluvial loam, that covers the gypsum quarries in new red sand-stone at Thiede, about four miles south-west of Brunswick. The remains lay heaped on each other, as represented in the plate, all within a space of 10 feet square. Among them were 11 tusks and 30 molar teeth of elephants (one 14 feet long), and various bones and teeth of elephant, rhinoceros, horse, ox, and stag.

In the drawing, the letters and figures which are not placed on teeth or bones imply that they existed in the loam in the place immediately below the figures respectively.

PLATE XXV.

Fig. 1. Sectional view of the coast of Dorset, from Lyme Regis to the Isle of Portland, as seen from Lyme Regis; showing the manner in which the valleys are intersected at the point where they terminate in the present sea-shore. It is probable that a considerable portion of this coast has been worn away by the sea, and that the small clay valleys or combs, which are now abruptly truncated at their termination, were originally continued with a gradual slope to the water's edge. The form of these cliffs, and of those in fig. 2, as they appear when seen from a boat in passing along the coast, is represented in the sections by Mr. De la Beche, at Plate VIII. of the Geological Transactions, second Series, vol. i. part i.

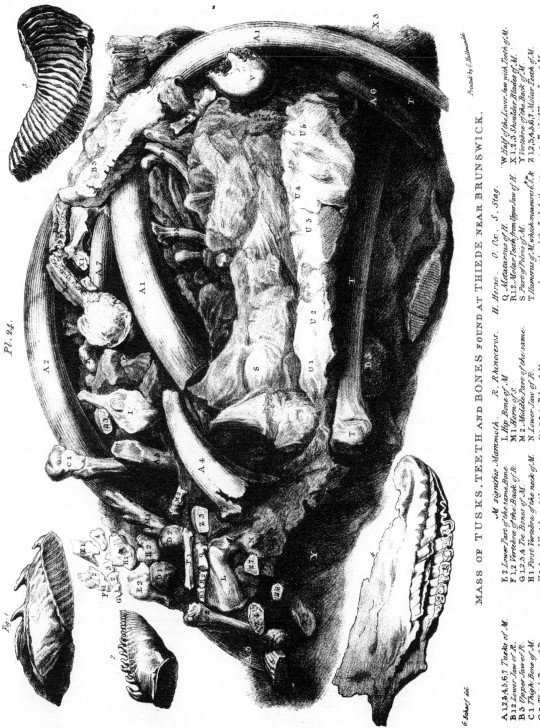

Pl. 24.

G. Scharf del.   Printed by C. Hullmandel.

MASS OF TUSKS, TEETH AND BONES FOUND AT THIEDE NEAR BRUNSWICK.

M signifies Mammoth.   R. Rhinoceros.   H. Horse.   O. Ox.   S. Stag.

A 1,2,3,4,5,6,7 Tusks of M.
B 1,2 Lower Jaw of R.
B 3 Upper Jaw of R.
C 1 Thigh Bone of M.
C 2 Thigh Bone of R.
D 1,2,3 Head Part of Thigh Bone of M.
E 1 Upper Part of Humerus of M.

E 2 Lower Part of the same Bone.
F 1,2 Vertebra of the Back of R.
G 1,2,3,4 The Bones of M.
H 1 First Vertebra of the neck of M.
H 2 Second Vertebra of the same.
I Lower Jaw of M.
K Patella of M.

L Hip Bone of M.
M 1 Horn of S.
M 2 Middle Part of the same.
N Lower Jaw of R.
O 1,2,3,4 Ribs of M.
P 1 Tibia of H.
P 2 Tibia of S.

Q Metatarsus of H.
R 1,2 Molar Teeth from Upper Jaw of H.
S Part of Pelvis of M.
T Humerus of M which measures 8, &c.
  may be considered the Scale for the rest.
U 1,2,3,4,5 Cervical Vertebra of M.
V Part of a Skull of M.

W Half of the Lower Jaw with Teeth of M.
X 1,2,3 Shoulder Blade of M.
Y Vertebra of the Back of M.
2. 1,2,3,4,5,6,7 Molar Teeth of M.
1. 2. Teeth of Upper Jaw of M.
3. Tooth of Lower Jaw of M.
4. Upper Jaw of R.

MAP of the VALE of the EVENLODE, THAMES, &c.

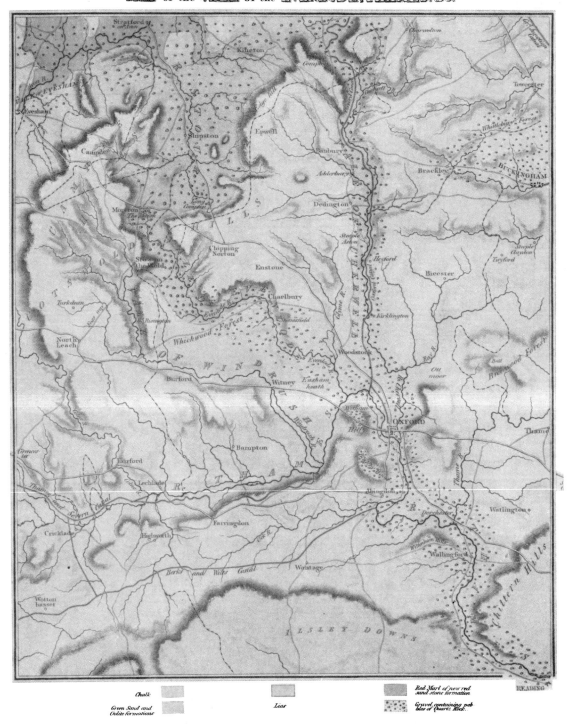

Chalk

Green Sand and Oolite formations

Lias

Red Marl of new red sand stone formation

Gravel containing pebbles of Quartz Rock.

Pl.26.

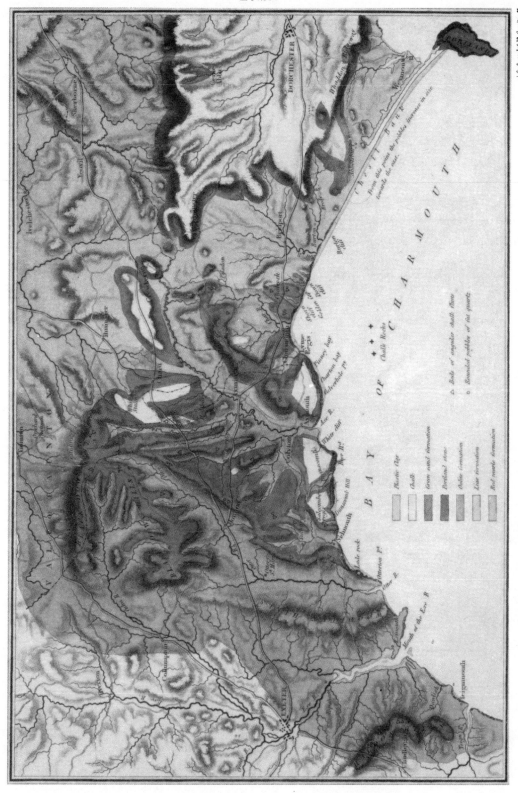

MAP of the COUNTRY adjacent to the COAST of DEVON and DORSET from TEIGNMOUTH to PORTLAND.

Pl. 25.

Nᵒ 1.

Section *of the* Coast *of* Dorset *from* Lyme Regis, *to the* Isle *of* Portland

Blue Lias.

Oolite.

Green sand

Drawn *by* H Cornish Esqʳᵉ

Section *of the* Coast *of* Devonshire *from* Sidmouth *to* Beerhead

G Scharf Lithog

Red Marl.

Green sand.

Chalk.

Beds of Chalk Flints.

Printed by C.Hullmandel

Portland Island.

Abbotsbury Castle.

Burton Cliff.

Bridport Harbour.

Down Cliffs.

Golden Cap.

Shorne Cliff.

Valley of Charmouth.

Black Ven.

Brandscombe Cliffs.

Little Combe Hill.

Dunscombe Hill.

Salcombe Hill.

Valley of Sidmouth

Peak Hill.

Fig. 2. Sectional view of the coast of Devonshire, from Sidmouth to Beer-head. The first comb or dry valley, on the east of Sidmouth, is abruptly truncated, like those represented in fig. 1.; the others terminate by a gradual slope towards the sea. The line of junction of the green-sand with the red marl is marked by the cessation of in-closures and of fertile soil, exactly at the point where the green-sand begins. The table-lands, that form the summits of these green-sand hills, are for the most part barren heaths, except where they are covered with diluvian gravel, or by a bed of unrolled chalk flints. This observation applies also to the green-sand summits in fig. 1, and to the table-lands composed of the same stratum, which stretch in-land from the coast to the flat summits of the Black Down Hills, in which this formation attains its highest elevation, overhanging with its escarpment the vale of Taunton. (See map at Plate XXVI.)

PLATE XXVI.

Map of the valleys which intersect the coast of Dorset and Devon.—The north-west angle, not being mentioned in the paper, is not coloured.

PLATE XXVII.

Map showing the manner in which the Lickey sand-stone peb-bles have been drifted from Warwickshire, through two low points in the escarpment of the oolite limestone at Moreton in Marsh, and on the north of Banbury; and been spread over the country along the valleys of the Evenlode, the Cherwell, and the Thames, and also on the north of Buckingham.

TABLE,

Shewing the principal Localities of the Antediluvian Animals mentioned in this Work.

	In caves or fissures.													In superficial loam or gravel.								
	Kirkdale.	Wirksworth.	Mendip.	Clifton.	Plymouth.	Crawley Rocks, near Swansea.	Paviland Caves, near Swansea.	District of Muggendorf.	District of the Hartz.	Fouvent, in France.	Sundwick, in Westphalia.	Köstritz, near Leipsig.	Gibraltar.	Oxford.	Brentford.	Walton, in Essex.	Lawford, near Rugby.	Thiede, near Brunswick.	Herzberg and Osterode.	Canstadt, near Stutgard.	Eichstadt, in Bavaria.	Val d'Arno, near Florence.
Hyæna	~~				~~	~~	~~	~~	~~	~~	~~	~~					~~		~~	~~	~~	~~
Tiger	~~				~~			~~	~~			~~	~~									~~
Bear	~~		~~		~~		~~	~~	~~		~~	~~										~~
Wolf	~~				~~		~~	~~	~~													
Fox	~~		~~		~~		~~	~~	~~													~~
Weasel	~~																			~~		
Elephant	~~	~~	~~			~~	~~	~~		~~				~~	~~	~~	~~	~~	~~	~~	~~	~~
Rhinoceros	~~	~~			~~		~~		~~		~~	~~		~~		~~	~~	~~	~~	~~		~~
Hippopotamus	~~														~~	~~						~~
Horse	~~		~~	~~	~~		~~			~~		~~	~~	~~		~~	~~	~~		~~		~~
Ox	~~	~~			~~	~~						~~	~~	~~	~~	~~	~~	~~		~~		~~
Deer	~~	~~	~~		~~	~~	~~				~~	~~	~~	~~	~~	~~	~~	~~		~~	~~	~~
Rabbit or Hare	~~												~~							~~		
Water-rat	~~												~~									
Mouse	~~												~~									
Birds	~~												~~				~~					
Glutton								~~	~~			~~										
Hog			~~												~~							~~
Mastodon																						~~
Tapir																						~~
Castor																						~~

INDEX.

A.

ABINGDON, pebbles of porphyry in gravel, 198.

———— bones of elephants and other animals in gravel at, 175.

———— quartzose pebbles in the valley of, and on the hills adjacent, 252.

Abscess, marks of, discovered on antediluvian wolf's bone by Mr. Cliff, 74.

Abury, Druidical temple built of grey wethers, 248.

Accidents, many must concur to the discovery of bones in caverns, 97.

Actual causes, their effects on lake Huron, 216.

———— began at the period usually assigned, 228.

———— cannot have produced diluvial phenomena, 227.

Adesberg, cave in Carniola, containing bears' bones, 161.

Adipocere, in human foot at Paviland, 88.

Ælian mentions hyænas' enmity to dogs, 23.

Africa, proofs of diluvial action in, 220.

Agates, obtained from diluvium, in Hindoostan, ibid.

Aikin, Mr. his account of gravel at Litchfield, 195.

Album græcum, calcareous excrement of hyænas in cave at Kirkdale, 20.

———— shows hyænas to have eaten bones, 37.

Allan, Mr. his account of osseous breccia at Nice, 151, 152.

Alluvium of two distinct eras, 185.

———— term applied too vaguely, 190.

———— recent, Mr. Bald's description of it, 186.

———— ancient, Mr. Bald's description of it, ibid.

Alps have been under water, 221.

——— effects of diluvial action in them enormous, 212.

Alpine limestone in Franconia, 124.

Alternation, none of stalagmite with beds of loam and pebbles in the caves, 110.

———— none of mud and stalagmite at Biels Hole, in the Hartz, 123.

America, proofs of diluvial action in, 215.

———— was inundated at the same time with Europe and Asia, 118.

———— North, pebbles of lead in stream works, 177.

———— South, proofs of diluvial action in it, 118.

Analogy of diluvial phenomena in all countries, 227.

Angular fragments, more abundant in fissures than in caves, 151.

———— in Mediterranean breccia, 150.

———— in caves, are of two eras, 143.

——— flints, on summits, in Devon and Dorset, 245.

o o

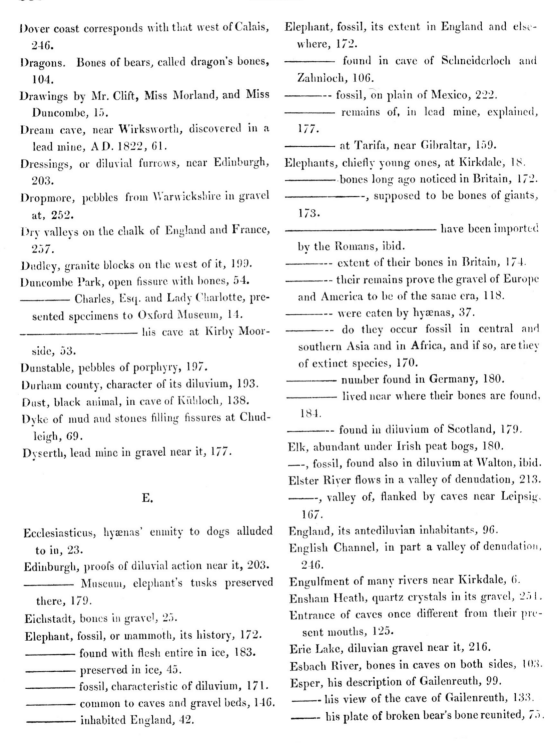

Q Q 2

THE END.

LONDON:
PRINTED BY THOMAS DAVISON, WHITEFRIARS.

History of Geology

An Arno Press Collection

Association of American Geologists and Naturalists. **Reports of the First, Second, and Third Meetings of the Association of American Geologists and Naturalists, at Philadelphia, in 1840 and 1841, and at Boston in 1842.** 1843

Bakewell, Robert. **An Introduction to Geology.** 1833

Buckland, William. **Reliquiae Diluvianae:** Or, Observations on the Organic Remains Contained in Caves, Fissures, and Diluvial Gravel. 1823

Clarke, John M[ason]. **James Hall of Albany:** Geologist and Palaeontologist, 1811-1898. 1923

Cleaveland, Parker. **An Elementary Treatise on Mineralogy and Geology.** 1816

Clinton, DeWitt. **An Introductory Discourse:** Delivered Before the Literary and Philosophical Society of New-York on the Fourth of May, 1814. 1815

Conybeare, W. D. and William Phillips. **Outlines of the Geology of England and Wales.** 1822

Cuvier, [Georges]. **Essay on the Theory of the Earth.** Translated by Robert Kerr. 1817

Davison, Charles. **The Founders of Seismology.** 1927

Gilbert, G[rove] K[arl]. **Report on the Geology of the Henry Mountains.** 1877

Greenough, G[eorge] B[ellas]. **A Critical Examination of the First Principles of Geology.** 1819

Hooke, Robert. **Lectures and Discourses of Earthquakes and Subterraneous Eruptions.** 1705

Kirwan, Richard. **Geological Essays.** 1799

Lambrecht, K. and W. and A. Quenstedt. **Palaeontologi:** Catalogus Bio-Bibliographicus. 1938

Lyell, Charles. **Charles Lyell on North American Geology.** Edited by Hubert C. Skinner. 1977

Lyell, Charles. **Travels in North America in the Years 1841-2.** Two vols. in one. 1845

Marcou, Jules. **Jules Marcou on the Taconic System in North America.** Edited by Hubert C. Skinner. 1977

Mariotte, [Edmé]. **The Motion of Water and Other Fluids.** Translated by J. T. Desaguliers. 1718

Merrill, George P., editor. **Contributions to a History of American State Geological and Natural History Surveys.** 1920

Miller, Hugh. **The Old Red Sandstone.** 1857

Moore, N[athaniel] F. **Ancient Mineralogy.** 1834

[Murray, John]. **A Comparative View of the Huttonian and Neptunian Systems of Geology.** 1802

Parkinson, James. **Organic Remains of a Former World.** Three vols. 1833

Phillips, John. **Memoirs of William Smith, LL.D.** 1844

Phillips, William. **An Outline of Mineralogy and Geology.** 1816

Ray, John. **Three Physico-Theological Discourses.** 1713

Scrope, G[eorge] Poulett. **The Geology and Extinct Volcanos of Central France.** 1858

Sherley, Thomas. **A Philosophical Essay.** 1672

Thomassy, [Marie Joseph] R[aymond]. **Géologie pratique de la Louisiane.** 1860

Warren, Erasmus. **Geologia:** Or a Discourse Concerning the Earth Before the Deluge. 1690

Webster, John. **Metallographia:** Or, an History of Metals. 1671

Whiston, William. **A New Theory of the Earth.** 1696

White, George W. **Essays on History of Geology.** 1977

Whitehurst, John. **An Inquiry into the Original State and Formation of the Earth.** 1786

Woodward, Horace B. **History of Geology.** 1911

Woodward, Horace B. **The History of the Geological Society of London.** 1907

Woodward, John. **An Essay Toward a Natural History of the Earth.** 1695